Two Manuals Combined

Military Intelligence

And

Electronic Warfare in Operations

Department of Defense

*FM 2-0, C1

Change Number 1

Headquarters
Department of the Army
Washington, DC, 17 May 2004

Intelligence

Contents

		Page
	PREFACE	v
PART ONE	**INTELLIGENCE IN THE OPERATIONAL ENVIRONMENT**	
Chapter 1	INTELLIGENCE AND THE OPERATIONAL ENVIRONMENT	1-1
	Role of Intelligence	1-1
	The Intelligence <u>Warfighting Function</u>	1-2
	Intelligence Tasks (METL)	1-3
	The Operational Environment	<u>1-20</u>
	The Intelligence Process	<u>1-22</u>
	Intelligence Disciplines	<u>1-25</u>
	Unified Action	1-27
	The Nature of Land Operations	1-28
	Force Projection Operations	1-29
Chapter 2	INTELLIGENCE AND UNIFIED ACTION	2-1
	Unified Action	2-1
	The Levels of War	2-1
	Categories of Intelligence	2-5
	Intelligence Community	2-7
	Unified Action Intelligence Operations	2-12
	Intelligence Reach	2-19

Distribution Restriction: Approved for public release; distribution is unlimited.

---

*This publication supersedes FM 34-1, 27 September 1994.

11 September 2008

i

FM 2-0, C1

PART TWO	INTELLIGENCE IN FULL SPECTRUM OPERATIONS	
Chapter 3	FUNDAMENTALS IN FULL SPECTRUM OPERATIONS	3-1
	Full Spectrum Operations	3-1
	Elements of Combat Power	3-3
	The Foundations of Army Operations	3-4
	The Operational Framework	3-4
	Army Capabilities	3-7
Chapter 4	INTELLIGENCE PROCESS IN FULL SPECTRUM OPERATIONS	4-1
	The Intelligence Process	4-1
	Plan	4-3
	Prepare	4-4
	Collect	4-8
	Process	4-10
	Produce	4-10
	Common Intelligence Process Tasks	4-11
PART THREE	MILITARY INTELLIGENCE DISCIPLINES	
Chapter 5	ALL-SOURCE INTELLIGENCE	5-1
	Definition	5-1
	Role	5-1
	Fundamentals	5-2
Chapter 6	HUMAN INTELLIGENCE	6-1
	Definition	6-1
	Role	6-1
	HUMINT Functions	6-1
	Operational Employment	6-4
	HUMINT Equipment	6-7
	Integration of Linguists	6-9
	Battle Hand-Off	6-10
	Organization	6-10
Chapter 7	IMAGERY INTELLIGENCE	7-1
	Definition	7-1
	Role	7-1
	Fundamentals	7-1

Chapter 8	SIGNALS INTELLIGENCE	8-1
	Definition	8-1
	Role	8-1
	Fundamentals	8-1
Chapter 9	MEASUREMENT AND SIGNATURES INTELLIGENCE	9-1
	Definition	9-1
	Role	9-3
	Fundamentals	9-4
Chapter 10	TECHNICAL INTELLIGENCE	10-1
	Definition	10-1
	Role	10-1
	Fundamentals	10-1
Chapter 11	COUNTERINTELLIGENCE	11-1
	Definition	11-1
	Role	11-1
	Counterintelligence Functions	11-2
	Operational Employment	11-6
	Counterintelligence Equipment	11-9
	Integration of Linguists	11-11
	Battle Hand-Off	11-12
	Organization	11-12
Appendix A	INTELLIGENCE AND INFORMATION OPERATIONS	A-1
	The Information Environment	A-1
	The Commander and Information	A-1
	Information Superiority	A-1
Appendix B	LINGUIST SUPPORT	B-1
	Role of Linguists	B-1
	Linguistic Support Categories	B-1
	Determining Linguist Requirements	B-1
	Planning and Managing Linguist Support	B-2
	Sources of Linguists	B-8
	Evaluating Linguist Proficiency	B-10
	Sustaining Military Linguist Proficiency	B-11

GLOSSARY	Glossary-1
BIBLIOGRAPHY	Bibliography-1
INDEX	Index-1

Preface

FM 2-0 is the Army's keystone manual for military intelligence (MI) doctrine. It describes—
- The fundamentals of intelligence operations.
- The operational environment (OE).
- Intelligence in unified action.
- The Intelligence Battlefield Operating System (BOS).
- Intelligence considerations in strategic readiness.
- The intelligence process.
- MI roles and functions within the context of Army operations.

This manual conforms to the overarching doctrinal precepts presented in FM 3-0.

This manual provides doctrinal guidance for the Intelligence BOS actions in support of commanders and staffs. It also serves as a reference for personnel who are developing doctrine; tactics, techniques, and procedures (TTP); materiel and force structure; and institutional and unit training for intelligence operations.

This manual provides MI guidance for all commanders, staffs, trainers, and MI personnel at all echelons. It forms the foundation for MI and the Intelligence BOS doctrine development, and applies equally to the Active Component (AC), United States Army Reserve (USAR), and Army National Guard (ARNG). It is also intended for commanders and staffs of joint and combined commands, US Naval and Marine Forces, units of the US Air Force, and the military forces of multinational partners.

Headquarters, US Army Training and Doctrine Command is the proponent for this publication. The preparing agency is the US Army Intelligence Center and School. Send written comments and recommendations on DA Form 2028 (Recommended Changes to Publications and Blank Forms) directly to: Commander, ATZS-FDT-D (FM 2-0), 550 Cibeque Street, Fort Huachuca, AZ 85613-7017. Send comments and recommendations by e-mail to *ATZS-FDC-D@hua.army.mil*. Follow the DA Form 2028 format or submit an electronic DA Form 2028.

Unless otherwise stated, masculine nouns and pronouns do not refer exclusively to men.

This manual contains Army tactical task (ART) description taken verbatim from the *Army Universal Task List* (AUTL). ART task descriptions are followed by a reference to FM 7-15 in parentheses and the ART number; for example (See FM 7-15, ART 5.1.1.).

PART ONE

Intelligence in the Operational Environment

Part One discusses MI's role in peace, conflict, and war. Supporting the warfighter with effective intelligence is the primary focus of Military Intelligence. Intelligence provides commanders and decisionmakers with the requisite information facilitating their situational understanding so that they may successfully accomplish their missions in full spectrum operations.

Chapter 1 describes the operational environment and the roles of MI within the operational environment. It introduces the Intelligence warfighting function, the intelligence tasks, and the intelligence process, which are the mechanisms through which MI supports the warfighter. This chapter also introduces the intelligence disciplines, which are explained in detail in Part Three of this manual.

Chapter 2 describes the interaction of MI within the nation's intelligence community structure, providing an overview of the intelligence community at the national level and the unified action level—joint, multinational, and interagency aspects of full spectrum operations. This chapter also discusses the concepts and components of intelligence reach.

Chapter 1

Intelligence and the Operational Environment

ROLE OF INTELLIGENCE

1-1. The commander requires intelligence about the enemy and the area of operations (AO) prior to engaging in operations in order to effectively execute battles, engagements, and other missions across the full spectrum of operations. Intelligence assists the commander in visualizing the AO, organizing the forces, and controlling operations to achieve the desired tactical objectives or end-state. Intelligence supports force protection (FP) by alerting the commander to emerging threats and assisting in security operations.

1-2. The unit may need to deal with multiple threats. The commander must understand how current and potential enemies organize, equip, train, employ, and control their forces. Intelligence provides an understanding of the enemy, which assists in planning, preparing, and executing military operations. The commander must also understand the operational environment and its effects on both friendly and enemy operations. The commander receives mission-oriented intelligence on enemy forces and the

FM 2-0, C1

AO from the G2/S2. The G2/S2 depends upon the intelligence, surveillance, and reconnaissance (ISR) effort to collect and provide information on the enemy and <u>AO</u>.

1-3. One of the most significant contributions that intelligence personnel can accomplish is to accurately predict future enemy events. Although this is an extremely difficult task, predictive intelligence enables the commander and staff to anticipate key enemy events or reactions and develop corresponding plans or counteractions. <u>Intelligence analysis conclusions, assessments, or products that attempt to define, describe, present, or portray the future situation or conditions of the enemy, terrain, weather, and civil considerations.</u>

1-4. The most important purpose of intelligence is to influence decisionmaking. Commanders must receive the intelligence, understand it (because it is tailored to the commander's requirements), believe it, and act on it. Through this doctrinal concept, intelligence drives operations.

THE INTELLIGENCE WARFIGHTING FUNCTION

1-5. *The intelligence warfighting function is the related tasks and systems that facilitate understanding of the operational environment.* (FM 3-0) It includes tasks associated with intelligence, surveillance, and reconnaissance (ISR) operations, and is driven by the commander. Intelligence is more than just collection. It is a continuous process that involves analyzing information from all sources and conducting operations to develop the situation. The intelligence warfighting function includes the following tasks:

- Support to force generation.
- Support to situational understanding.
- Perform intelligence, surveillance, and reconnaissance.
- Support to targeting and information operations.

1-6. The intelligence warfighting function is one of six warfighting functions—movement and maneuver, intelligence, fires, sustainment, command and control, and protection. *A warfighting function is a group of tasks and systems (people, organizations, information, and processes) united by a common purpose that commanders use to accomplish missions and training objectives* (FM 3-0). (See FM 3-0, chapter 4, for a detailed discussion of the warfighting functions.) The intelligence warfighting function is a flexible force of personnel, organizations, and equipment that, individually or collectively, provide commanders with the timely, relevant, and accurate intelligence required to visualize the battlefield, assess the situation, and direct military actions. Additionally, the intelligence warfighting function is—

- A complex system that operates worldwide, from "mud-to-space," in support of an operation, to include the ability to leverage theater and national capabilities.
- Requires cooperation and division of labor internally, higher, lower, adjacent, and across components and the coalition.

1-7. The intelligence warfighting function not only includes assets within the MI branch but also includes the assets of all branches or warfighting functions that conduct intelligence warfighting function tasks. Every Soldier, as a part of a small unit, is a potential information collector and an essential component to help reach situational understanding. Each Soldier develops a special level of awareness simply due to exposure to events occurring in the AO and has the opportunity to collect and report information by observation and interaction with the population (see paragraph 3-3).

1-8. Planning and executing military operations will require intelligence regarding the threat (traditional, irregular, catastrophic, and disruptive) and the AO. The intelligence warfighting function generates intelligence and intelligence products that portray the enemy and aspects of the environment. These intelligence products enable the commander to identify potential courses of action (COAs), plan operations, employ forces effectively, employ effective tactics and techniques, and implement protection.

1-9. The intelligence warfighting function is always engaged in supporting the commander in offensive, defensive, stability, and, when directed, civil support operations. Intelligence provides products that are timely, relevant, accurate, and predictive. Hard training, thorough planning, meticulous preparation, and aggressive execution posture the Army for success. In the current environment we must maintain intelligence readiness to support operations on "no notice." This support is comprehensive and reaches across full spectrum operations and levels of war to produce the intelligence required to successfully accomplish the mission. A combination of space, aerial, seaborne, and ground-based systems provide the most comprehensive intelligence possible. During force projection operations, the intelligence warfighting function supports the commander with accurate and responsive intelligence from predeployment through redeployment.

1-10. The intelligence warfighting function architecture provides specific intelligence and communications structures at each echelon from the national level through the tactical level. These structures include intelligence organizations, systems, and procedures for collecting, processing, analyzing, and delivering intelligence and other critical information in a useable form to those who need it, when they need it. Effective communications connectivity and automation are essential components of this architecture.

INTELLIGENCE TASKS (METL)

1-11. The personnel and organizations within the Intelligence <u>warfighting function</u> conduct four primary intelligence tasks that facilitate the commander's visualization and understanding of the threat and the AO. These tasks are interactive and often take place simultaneously. (Refer to FM 7-15 for the complete subordinate task listing.) Figure 1-1 shows these tasks tailored to the commander's needs.

FM 2-0, C1

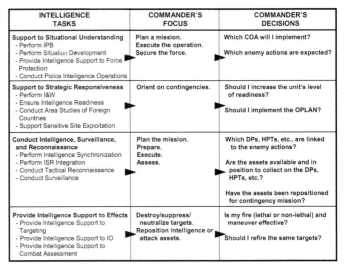

Figure 1-1. Intelligence Tailored to the Commander's Needs.

SUPPORT TO SITUATIONAL UNDERSTANDING

1-12. This task centers on providing information and intelligence to the commander, which facilitates the commander's understanding of the enemy and the environment. It supports the command's ability to make sound decisions. Support to situational understanding comprises four subtasks: perform intelligence preparation of the battlefield (IPB), perform situation development, provide intelligence support to FP, and conduct police intelligence operations.

Perform Intelligence Preparation of the Battlefield

1-13. The G2/S2 is the staff proponent for IPB. IPB is the staff planning activity undertaken by the entire staff to define and understand the battlefield and the options it presents to friendly and threat forces. IPB includes input from the whole staff. There is only one IPB in each headquarters with inputs from all affected staff cells; they are not separate warfighting function or staff section IPBs throughout the headquarters. It is a systematic process of analyzing and visualizing the operational environment in a specific geographic area for a specific mission or in anticipation of a specific mission. By applying IPB, the commander and staff gain the information necessary to selectively apply and maximize combat power at critical points in time and space. IPB is most effective when it integrates each staff element's expertise into the final products. To conduct effective IPB, the G2/S2 must—

- Produce IPB products that support the staff's preparation of estimates and the military decision-making process (MDMP).

- Identify characteristics of the AO, including the information environment, that will influence friendly and threat operations.
- Establish the area of interest (AOI) in accordance with the commander's guidance.
- Identify gaps in current intelligence holdings.
- Determine multiple enemy COAs (ECOAs) by employing predictive analysis techniques to anticipate future enemy actions, capabilities, or situations.
- Establish a database that encompasses all relevant data sets within and related to the <u>operational environment.</u>
- Determine the enemy order of battle (OB), doctrine, and TTP. Identify any patterns in enemy behavior or activities.
- Accurately identify and report hazards within the AO, including the medical threat and toxic industrial material (TIM).
- Accurately identify threat capabilities, high-value targets (HVTs), and threat models.
- Integrate IPB information into COA analysis and the MDMP.
- Update IPB products as information becomes available.

Perform Situation Development

1-14. Situation development is a process for analyzing information and producing current intelligence about the enemy and environment during operations. The process helps the intelligence officer recognize and interpret the indicators of enemy intentions, objectives, combat effectiveness, and potential ECOAs. Situation development—

- Confirms or denies threat COAs.
- Provides threat locations.
- Explains what the threat is doing in relation to the friendly force operations.
- Provides an estimate of threat combat effectiveness.

1-15. Through situation development, the intelligence officer is able to quickly identify information gaps, and explain enemy activities in relation to the unit's own operations, thereby assisting the commander in gaining situational understanding. Situation development helps the commander make decisions and execute branches and sequels. This reduces risk and uncertainty in the execution of the plan. The intelligence officer maintains, presents, and disseminates the results of situation development through intelligence input to the common operational picture (COP) and other intelligence products.

Provide Intelligence Support to Force Protection

1-16. Provide intelligence in support of protecting the tactical force's fighting potential so that it can be applied at the appropriate time and place. This task includes the measures the force takes to remain viable and functional by protecting itself from the effects of or recovery from enemy activities. FP consists of those actions taken to prevent or mitigate hostile actions against Department of Defense (DOD) personnel (to include Department of Army

[DA] civilians, contractors, uniformed personnel, and family members), resources, facilities, and critical information. These actions—
- Conserve the force's fighting potential for application at the decisive time and place.
- Incorporate coordinated and synchronized offensive and defensive measures.
- Facilitate the effective employment of the joint force while degrading the capabilities of and opportunities for the threat.

1-17. Intelligence to FP consists of monitoring and reporting the activities, intentions, and capabilities of adversarial groups and determining their possible COAs. Detecting the adversary's methods in today's OE requires a higher level of situational understanding, informed by current and precise intelligence. This type of threat drives the need for predictive intelligence based on analysis of focused information from intelligence, law enforcement, and security activities.

1-18. Intelligence analysis in support of FP employs analytical methodologies and tools to provide situational understanding and to predict the adversary's actions. Modified or standard time-event charts, association matrices, activity matrices, link diagrams, and overlays are beneficial in monitoring the actions of the adversary. Overlays may include (but are not limited to) threat training camps, organizations, finances, personalities, industrial sites, information systems, decisionmaking infrastructures, specific activities, and locations of previous attacks.

Conduct Police Intelligence Operations

1-19. Police intelligence operations (PIO) are a military police (MP) function that supports, enhances, and contributes to the commander's force protection program, COP, and situational understanding. The PIO function ensures that information collected during the conduct of other MP functions—maneuver and mobility support, area security, law and order, and internment and resettlement—is provided as input to the intelligence collection effort and turned into action or reports. PIO has three components: (See FM 7-15, ART 1.1.4)
- Collect police information.
- Conduct Police Information Assessment Program (PIAP).
- Develop police intelligence products.

Notes:

US Code, Executive Orders, DOD Directives, and Army Regulations contain specific guidance regarding prohibition on the collection of intelligence information on US citizens, US corporations, and resident aliens. These laws and regulations include criminal penalties for their violation. Any PIO directed against US citizens should undergo competent legal review prior to their initiation.

The inclusion of the PIO task branch in the Intelligence warfighting function does not change the intelligence process described in this manual.

SUPPORT TO STRATEGIC RESPONSIVENESS

1-20. Intelligence support to strategic responsiveness supports staff planning and preparation by defining the full spectrum of threats, forecasting future threats, and forewarning the commander of enemy actions and intentions. Support to strategic responsiveness consists of four subtasks: Perform I&W, ensure intelligence readiness, conduct area studies of foreign countries, and support sensitive site exploitation. (See FM 7-15, ART 1.2)

Perform Indications and Warnings

1-21. This activity provides the commander with forewarning of enemy actions or intentions; the imminence of threat actions. The intelligence officer develops I&W in order to rapidly alert the commander of events or activities that would change the basic nature of the operations. It enables the commander to quickly reorient the force to unexpected contingencies and shape the battlefield. (See FM 7-15, ART 1.2.1)

1-22. The G2/S2 at the operational and strategic levels develops I&W in order to rapidly alert the commander of events or activities that would change the basic nature of the operations so the commander can initiate the appropriate action in a timely manner. I&W reduce the risk of enemy actions that are counter to planning assumptions. I&W enable the commander to quickly reorient the force to unexpected events and to shape the battlefield by manipulating enemy activities.

Ensure Intelligence Readiness

1-23. Intelligence readiness operations support contingency planning and preparation by developing baseline knowledge of multiple potential threats and operational environments. These operations and related intelligence training activities engage the Intelligence <u>warfighting function</u> to respond effectively to the commander's contingency planning intelligence requirements. (See FM 7-15, ART 1.2.2) While still in garrison, intelligence defines the full spectrum of threats and forecasts future threats and dangers. Intelligence readiness operations accomplish the following:

- Provide intelligence to support contingency-based training and staff planning.
- Identify, consider, and evaluate all potential threats to the entire unit.
- Provide a broad understanding of the operational environment of the contingency area, which is developed through continuous exchange of information and intelligence with higher echelon and joint intelligence organizations.

Conduct Area Studies of Foreign Countries

1-24. Study and understand the cultural, social, political, religious, and moral beliefs and attitudes of allied, host nation (HN), or indigenous forces to assist in accomplishing goals and objectives. (See FM 7-15, ART 1.2.3)

Note: The inclusion of this task does not change the support to strategic responsiveness provided by MI organizations described in this manual.

FM 2-0, C1

Support Sensitive Site Exploitation

1-25. Sensitive site exploitation consists of a related series of activities inside a sensitive site captured from an adversary. A sensitive site is a designated, geographically limited area with special military, diplomatic, economic, or information sensitivity for the United States. This includes factories with technical data on enemy weapon systems, war crimes sites, critical hostile government facilities, areas suspected of containing persons of high rank in a hostile government or organization, terrorist money laundering, and document storage areas for secret police forces. These activities exploit personnel, documents, electronic data, and material captured at the site, while neutralizing any threat posed by the site or its contents. While the physical process of exploiting the sensitive site begins at the site itself, full exploitation may involve teams of experts located around the world. (See FM 7-15, ART 1.2.4)

Note: The inclusion of this task does not change the support to strategic responsiveness provided by MI organizations described in this manual.

CONDUCT INTELLIGENCE, SURVEILLANCE, AND RECONNAISSANCE

1-26. With staff participation, the intelligence officer synchronizes intelligence support to the ISR effort by focusing the collection, processing, analysis, and intelligence products on the critical needs of the commander. The operations officer, in coordination with the intelligence officer, tasks and directs the available ISR assets to answer the commander's critical information requirements (CCIRs). Through various detection methods and systematic observation, reconnaissance and surveillance obtains the required information. A continuous process, this task has four subtasks: perform intelligence synchronization, perform ISR integration, conduct tactical reconnaissance, and conduct surveillance. (See FM 7-15, ART 1.3, and FM 3-90).

Perform Intelligence Synchronization

1-27. The intelligence officer, with staff participation, synchronizes the entire collection effort to include all assets the commander controls, assets of lateral units and higher echelon units and organizations, and intelligence reach to answer the commander's priority intelligence requirements (PIRs) and information requirements (IRs). (See FM 7-15, ART 1.3.1)

1-28. The intelligence officer, with staff participation, supports the G3/S3 in orchestrating the entire ISR effort to include all assets the commander controls, assets of lateral units and higher echelon units and organizations, and intelligence reach to answer the CCIRs (PIRs and friendly force information requirements [FFIRs]) and other intelligence requirements. Intelligence synchronization activities include the following:

- Conducting requirements management (RM): anticipate, develop, analyze, validate, and prioritize intelligence requirements. Recommend PIRs to the commander. Manage the commander's intelligence requirements, requests for information (RFIs) from subordinate and lateral organizations, and tasks from higher headquarters. Eliminate satisfied requirements and add new requirements as necessary.

- Developing indicators.
- Developing specific IRs (specific information requirements [SIRs]).
- Converting the SIRs into intelligence tasks or ISR tasks. (See Figure 1-2 for the ISR task development process.) The S2/G2 assigns intelligence production and reach tasks to subordinate intelligence elements or personnel, submits RFIs to higher and lateral echelons, and coordinates with (or assists) the G3/S3 to develop and assign ISR tasks.
- Comparing the ISR tasks to the capabilities and limitations of the available ISR assets (in coordination with the operations officer).
- Forwarding SIRs that cannot be answered by available assets to higher or lateral organizations as RFIs.
- Assessing collection asset reporting and intelligence production to evaluate the effectiveness of the ISR effort.
- Maintaining situational understanding to identify gaps in coverage and to identify the need to cue or redirect ISR assets.
- Updating the intelligence synchronization plan. The G2/S2 manages and updates the intelligence synchronization plan as PIRs are answered and new requirements arise.

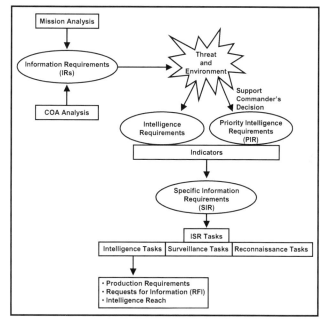

Figure 1-2. ISR Task Development Process.

1-29. **Intelligence Synchronization Considerations.** The G2/S2 generally follows six considerations in planning intelligence synchronization and ISR activities: anticipate, integrate, prioritize, balance, control, and reach. Refer to FM 2-01 for more information regarding intelligence synchronization.

- **Anticipate.** The intelligence staff must recognize when and where to shift collection or identify new intelligence requirements. The overall intent of this principle is to identify a new or adjust an existing requirement and present it to the commander for approval before waiting for the commander or staff to identify it.
- **Integrate.** The intelligence staff must be fully and continuously integrated into the unit's orders production and planning activities to ensure early identification of intelligence requirements. Early and continuous consideration of collection factors enhances the unit's ability to direct collection assets in a timely manner, ensures thorough planning, and increases flexibility in selecting assets.
- **Prioritize.** Prioritize each intelligence requirement based on its importance in supporting the commander's intent and decisions. Prioritization, based on the commander's guidance and the current situation, ensures that limited ISR assets and resources are directed against the most critical requirements.
- **Balance.** ISR capabilities complement each other. The intelligence staff should resist favoring or becoming too reliant on a particular unit, discipline, or system. Balance is simply planning redundancy, when required, eliminating redundancy when not desired, and ensuring an appropriate mix of ISR assets or types. The intelligence synchronization matrix (ISM) is useful in determining or evaluating balance.
- **Control.** To ensure timely and effective responses to intelligence requirements, a unit should first use ISR assets it controls. These assets usually are more responsive to their respective commander and also serve to lessen the burden on the ISR assets of other units, agencies, and organizations.
- **Reach.** Although usually not as responsive as a unit's own assets, intelligence reach may be the only way to satisfy an intelligence requirement. If at all possible, one should not depend solely on intelligence reach to answer a PIR.

1-30. **Develop IRs.** The intelligence staff develops a prioritized list of what information needs to be collected and produced into intelligence. Additionally, the intelligence staff dynamically updates and adjusts those requirements in response to mission adjustments and changes. This list is placed against a latest time intelligence is of value to ensure intelligence and information are reported to meet operational requirements. (See FM 7-15, ART 1.3.1) Figure 1-3 shows a comparison of IR, PIR, and intelligence requirements.

FM 2-0, C1

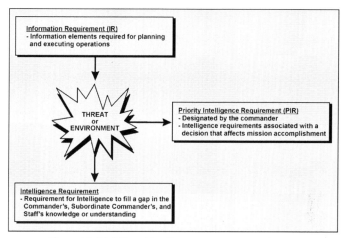

Figure 1-3. IR, PIR, and Intelligence Requirements Comparison.

1-31. An effective discussion of ISR has to include an understanding of the CCIRs. The CCIRs are elements of information required by commanders that directly affect decisionmaking and dictate the successful execution of military operations. The commander decides what information is critical based on the commander's experience, the mission, the higher commander's intent, and the staff's input (initial IPB, information, intelligence, and recommendations). Refer to FM 3-0 for more information regarding CCIRs.

1-32. Based on the CCIRs, two types of supporting IRs are generated: PIRs and FFIRs. However, commanders may determine that they need to know whether one or more essential elements of friendly information (EEFI) have been compromised or that the enemy is collecting against a designated EEFI. In those cases, commanders may designate that question as one of their CCIRs. Figure 1-4 shows the CCIR composition.

1-33. IRs are all of the information elements required by the commander and staff for the successful planning and execution of operations; that is, all elements necessary to address the factors of mission, enemy, terrain and weather, troops and support available, time available, civil considerations (METT-TC). Vetting by the commander or a designated representative turns an IR into either a PIR or an intelligence requirement. IRs are developed during COA analysis based on the factors of METT-TC.

1-34. PIRs are those intelligence requirements for which a commander has an anticipated and stated priority during the task of planning and decisionmaking. PIRs are associated with a decision based on an enemy action or inaction or the AO that will affect the overall success of the commander's mission. **The commander designates intelligence requirements tied directly to decisions as CCIR (PIR and FFIR).**

11 September 2008   1-11

Answers to the PIRs help produce intelligence essential to the commander's situational understanding and decisionmaking. For information on PIR development, see FM 2-01.

1-35. The G2/S2 recommends to the commander those IRs produced during the MDMP that meet the criteria for PIR. They do not become CCIR (PIR and FFIR) until approved by the commander. Additionally, The commander may unilaterally designate PIRs. The IRs that are not designated by the commander as PIRs become intelligence requirements. The intelligence requirement is a gap in the command's knowledge or understanding of the AO or threat that the Intelligence warfighting function must fill.

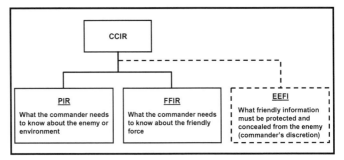

Figure 1-4. Commander's Critical Information Requirements Composition.

1-36. The G3/S3 then tasks the unit's assets to answer both the PIR and intelligence requirements through the ISR plan. PIR should—
- Ask only one question.
- Support a decision.
- Identify a specific fact, event, activity (or absence thereof) which can be collected.
- If linked to an ECOA, indicate an ECOA prior to, or as early as possible in, its implementation.
- Indicate the latest time the information is of value (LTIOV). The LTIOV is the absolute latest time the information can be used by the commander in making the decision the PIR supports. The LTIOV can be linked to time, an event, or a point in the battle or operation.

1-37. **Friendly Force IRs.** The staff also develops FFIRs which, when answered, provide friendly force information that the commander and staff need to achieve situational understanding and to make decisions.

1-38. **Essential Elements of Friendly Information.** EEFI establish information to protect, not information to obtain. However, commanders may determine that they need to know whether one or more EEFI have been compromised or that the enemy is collecting against a designated EEFI. In those cases, commanders may designate that question as one of their CCIRs, which generates PIRs and/or FFIRs. For example, a commander may determine that if the enemy discovers the location and movement of the friendly reserve, the operation is at risk. In this case, the location and movement of the friendly reserve are EEFI. The commander designates determining whether the enemy has discovered the location and movement of the friendly reserve as one of the CCIR. That CCIR, in turn, generates PIR and FFIR to support staff actions in determining whether the EEFI has been compromised.

1-39. **Develop the Intelligence Synchronization Plan.** The entire unit staff develops their IRs and determines how best to satisfy them. The staff uses reconnaissance and surveillance assets to collect information. The intelligence synchronization plan includes all assets that the operations officer can task or request and coordination mechanisms to ensure adequate coverage of the AOIs. (See FM 7-15, ART 1.3.1.2)

1-40. The intelligence synchronization plan, often presented in a matrix format as an ISM, aids in synchronizing the entire ISR effort with the overall operation and the commander's decisions and/or decision points (DPs). The intelligence synchronization plan is often produced in conjunction with the ISR plan. However, before performing intelligence synchronization and finalizing the intelligence synchronization plan, the G2/S2 must have—
- The CCIR (PIR and FFIR).
- A prioritized list of the remaining intelligence requirements.
- Evaluated ISR assets and resources.
- All of the assigned ISR tasks.

Perform ISR Integration

1-41. The operations officer, in coordination with the intelligence officer and other staff members, orchestrates the tasking and directing of available ISR assets to answer the CCIR. The operations officer, with input from the intelligence officer, develops tasks from the SIRs which coincide with the capabilities and limitations of the available ISR assets and the latest time information is of value (LTIOV). Intelligence requirements are identified, prioritized, and validated and an ISR plan is developed and synchronized with the scheme of maneuver. (See FM 7-15, ART 1.3.2, and FM 3-90)

1-42. The G3/S3, in coordination with the G2/S2 and other staff members, orchestrates the tasking and directing of available ISR assets to answer the CCIRs (PIRs and FFIRs) and IRs. The result of this process is the forming of the ISR plan. The ISR plan provides a list of all the ISR tasks to be accomplished. The G2/S2 and the G3/S3 develop tasks from the SIRs. These tasks are then assigned based on the capabilities and limitations of the available ISR assets and the LTIOV.

1-43. **Develop the ISR Plan.** The operations officer is responsible for developing the ISR plan. The entire unit staff analyzes each requirement to determine how best to satisfy it. The staff will receive orders and RFIs from both subordinate and adjacent units and higher headquarters. The ISR plan includes all assets that the operations officer can task or request and coordination mechanisms to ensure adequate coverage of the area of interest. (See FM 7-15, ART 1.3.2.1, and FM 3-90)

1-44. **The ISR Plan.** The ISR plan is usually produced as the ISR Annex to an OPORD (Annex L, Intelligence Surveillance, and Reconnaissance). Refer to FM 5-0 for specific and authoritative information on the ISR Annex. ISR is a continuous combined arms effort led by the operations and intelligence staffs in coordination with the entire staff that sets reconnaissance and surveillance in motion. The PIRs and other intelligence requirements drive the ISR effort. The commander takes every opportunity to improve situational understanding and the fidelity of the COP about the enemy and terrain through the deployment of ISR assets. Commanders integrate reconnaissance and surveillance to form an integrated ISR plan that capitalizes on their different capabilities. The ISR plan is often the most important part of providing information and intelligence that contributes to answering the CCIRs. For the G2/S2, an effective ISR plan is critical in answering the PIR. Also see FM 3-55 for more information on the ISR plan.

1-45. The ISR plan is not an MI-specific product—the G3/S3 is the staff proponent of the ISR plan— it is an integrated staff product executed by the unit at the direction of the commander. The G2/S2, however, must maintain situational understanding in order to recommend to the commander and G3/S3 changes or further development of the ISR plan. Based on the initial IPB and CCIRs, the staff—primarily the G2/S2—identifies gaps in the intelligence effort and develops an initial ISR plan based on available ISR assets. The G3/S3 turns this into an initial ISR Annex that tasks ISR assets as soon as possible to begin the collection effort.

1-46. The G3/S3, assisted by the G2/S2, uses the ISR plan to task and direct the available ISR assets to answer the CCIRs (PIRs and FFIRs) and intelligence requirements. Conversely, the staff revises the plan as other intelligence gaps are identified if the information is required to fulfill the CCIRs or in anticipation of future intelligence requirements. With staff participation, the G2/S2 intelligence officer synchronizes the ISR effort through a complementary product to the ISR plan—the intelligence synchronization plan.

1-47. **Execute and Update the ISR Plan.** The operations officer updates the ISR plan based on information received from the intelligence officer. The operations officer is the integrator and manager of the ISR effort through an integrated staff process and procedures. As PIRs are answered and new information requirements arise, the intelligence officer updates intelligence synchronization requirements and provides the new input to the operations officer who updates the ISR plan. He works closely with all staff elements to ensure the unit's organic collectors receive appropriate taskings. This ISR reflects an integrated collection strategy and employment, production and dissemination scheme that will effectively answer the commander's PIR. (See FM 7-15, ART 1.3.2.2, and FM 3-90)

Conduct Tactical Reconnaissance

1-48. To obtain, by visual observation or other detection methods, such as signals, imagery, measurement of signature or other technical characteristics, human interaction and other detection methods about the activities and resources of an enemy or potential enemy, or to secure data

concerning the meteorological, hydrographic, or geographic characteristics and the indigenous population of a particular area. This task includes the conduct of NBC reconnaissance and the tactical aspects of SOF special reconnaissance. The five subtasks are—
- Conduct a Zone Reconnaissance.
- Conduct an Area Reconnaissance.
- Conduct a Reconnaissance in Force.
- Conduct a Route Reconnaissance.
- Conduct a Reconnaissance Patrol.

Note: This task branch includes techniques by which ART 1.1.4.1 (Collect Police Information) may be performed. (FM 7-15)

1-49. Reconnaissance is a mission undertaken to obtain by visual observation or other detection methods, information about the activities and resources of an enemy or potential enemy, and about the meteorological, hydrographic, or geographic characteristics of an AO. MI personnel and organizations can conduct reconnaissance through obtaining information derived from signals, imagery, measurement of signatures, technical characteristics, human interaction, and other detection methods. When performing reconnaissance, it is important to—
- Orient the reconnaissance asset on the named area of interest (NAI) and/or reconnaissance objective in a timely manner.
- Report all information rapidly and accurately.
- Complete the reconnaissance mission not later than (NLT) the time specified in the order.
- Answer the requirement that prompted the reconnaissance task.

Conduct Surveillance

1-50. To systematically observe the airspace, surface, or subsurface areas, places, persons, or things in the AO by visual, aural (audio), electronic, photographic, or other means. Other means may include but are not limited to space-based systems, and using special NBC, artillery, engineer, SOF, and air defense equipment. (See FM 7-15, ART 1.3.4, and FM 3-90)

Note: This task is a technique by which ART 1.1.4.1 (Collect Police Information) may be performed.

1-51. Conducting surveillance is systematically observing the airspace, surface, or subsurface areas, places, persons, or things in the AO by visual, aural, electronic, photographic, or other means. Surveillance activities include—
- Orienting the surveillance asset on the NAI and/or the surveillance objective in a timely manner.
- Reporting all information rapidly and accurately.
- Completing the surveillance mission NLT the time specified in the order.
- Answering the requirement that prompted the surveillance task.

FM 2-0, C1

PROVIDE INTELLIGENCE SUPPORT TO EFFECTS

1-52. The task of providing the commander information and intelligence support for targeting of the threat's forces, threat organizations, units and systems through lethal and non-lethal fires to include electronic attack and information operations. This task includes three subtasks: provide intelligence support to targeting, provide intelligence support to information operations, and provide intelligence support to combat assessment. (See FM 7-15, ART 1.4)

Provide Intelligence Support to Targeting

1-53. The intelligence officer, supported by the entire staff, provides the commander information and intelligence support for targeting the threat's forces and systems through employment of direct, indirect lethal and non-lethal fires. It includes identification of threat capabilities and limitations. The Intelligence warfighting function plays a crucial role by providing intelligence throughout the steps of the targeting process: decide, detect, deliver, assess. Support to targeting includes identifying threat capabilities and limitations. The G2/S2 supports the unit's lethal and non-lethal targeting effort through target development, target acquisition (TA), electronic warfare (EW), and combat assessment. This task has two subtasks:

- *Provide Intelligence Support to Target Development* is the systematic analysis of the enemy forces and operations to determine HVTs, systems, and system components for potential attack through maneuver, fires, or information.
- *Provide Intelligence Support to Target Detection* establishes procedures for dissemination of targeting information. The targeting team develops the sensor/attack system matrix to determine the sensor required to detect and locate targets. The intelligence officer places the following requirements into the integrated ISR plan.
    - Requires reconnaissance and surveillance operations to identify, locate, and track high-payoff targets (HPTs) for delivery of lethal or non-lethal effects.
    - Includes employing fires, offensive IO, and other attack capabilities against enemy C2 systems as part of the unit's FS plan and IO objectives.

Provide Intelligence Support to Information Operations

1-54. IO are actions taken to affect adversary information, influence other's decisionmaking processes and information systems while protecting one's own information and information systems. Overall operational continuity and mission success requires close, mutual coordination and synchronization of intelligence plans and operations with IO elements and related activities. Refer to Appendix A, this manual, for more information. This task has three subordinate tasks: (See FM 7-15, ART 1.4.2)

- Provide Intelligence Support to Offensive IO.
- Provide Intelligence Support to Defensive IO.
- Provide Intelligence Support to Activities Related to IO.

Provide Intelligence Support to Combat Assessment

1-55. Combat assessment is the determination of the overall effectiveness of force employment during military operations. The objective of combat assessment is to identify recommendations for the course of military operations. It answers the question, "Were the objectives met by force employment?" Although the assessment is primarily an intelligence responsibility, it requires input from and coordination with operations and fire support staffs. Combat assessment consists of conducting physical damage, functional damage, and target system assessments. (See FM 7-15, ART 1.4.3)

1-56. The staff determines how combat assessment relates to a specific target by conducting physical damage, functional damage, and target system assessments.

- **Conduct Physical Damage Assessment (PDA).** A PDA is an estimate of the extent of physical damage to a target based upon observed or interpreted damage. This post-attack target analysis is a coordinated effort among all units.
- **Conduct Functional Damage Assessment (FDA).** The FDA estimates the remaining functional or operational capability of a targeted facility or object. The staff bases FDA on the assessed physical damage and estimates of the threat's ability to recuperate (to include the time required to resume normal operations). Multiple echelons typically conduct this all-source analysis. The targeting or combat assessment cell integrates the initial target analyses with other sources, including intelligence, and then compares the original objective with the current status of the target to determine if the objective has been met.
- **Conduct Target System Assessment (TSA).** The TSA is an estimate of the overall impact of force employment against an adversary's target system. The unit, supported by higher echelon assets, normally conducts this assessment. The analyst combines all combat assessment reporting on functional damage to targets within a target system and assesses the overall impact on that system's capabilities. This process lays the groundwork for future recommendations for military operations in support of operational objectives.

1-57. **Munitions Effects Assessment (MEA).** MEA takes place concurrently and interactively with combat assessment since the same signatures used to determine the level of physical damage also give clues to munitions effectiveness. MEA is primarily the responsibility of operations and FS personnel, with input from the G2/S2. After the same weapon is used to attack several targets of a specific type, MEA should be accomplished to evaluate weapon performance. MEA analysts seek to identify through systematic trend analysis any deficiencies in weapon system and munitions performance or combat tactics by answering the question, "Did the weapons employed perform as expected?" Using a variety of input (targeting analysts, imagery analysts, structural engineers, and mission planners) analysts prepare a report assessing munitions performance. If combat troops capture attacked targets, it is then possible to collect detailed information on the target. Reports should detail weapon performance against specified target types. This information could have a crucial impact on future operations.

1-58. **Re-attack Recommendation.** Re-attack recommendations follow directly from both battle damage assessment (BDA) and MEA efforts. Basically re-attack recommendations answer the question, "Have we achieved the desired effects against our targeted objectives?" Evolving objectives, target selection, timing, tactics, weapons, vulnerabilities, and munitions are all factors in the new recommendations, combining both operations and intelligence functions.

1-59. **Targeting Meeting.** The role of the G2/S2 in targeting meetings is critical to ensuring intelligence supports the targeting process. The G2/S2 must come to the targeting meeting prepared. Additionally, the G2/S2 must understand the steps of the targeting process and tailor the intelligence products and G2/S2 participation according to the targeting process steps of: decide, detect, deliver, and assess.

1-60. There are two major areas in which the G2/S2 must prepare in order to support the targeting meeting.

- **Analysis:** Determine what the enemy is doing (current situation) and anticipated threat models; ECOA situation templates, ECOA sketches and statements, HVTs, and other associated products.
- **IPB Products:** Updated from the initial IPB effort, including analytical results (see above bullet), and tailored to the targeting requirements.

1-61. **Decide.** At the targeting meeting, the G2/S2 should be prepared to do the following:

- Brief the current and projected future enemy situation with an event template.
- Brief any combat assessments.
- Brief HVTs for potential selection of HPTs.
- Brief the current PIR portion of CCIRs, the PIRs of the CCIRs for the next phase or in accordance with the same timeframe of the targeting meeting.
- Brief the current ISM or intelligence synchronization plan.
- Participate in wargaming.
- Refine the initial ISR plan (in conjunction with the FS and G3/S3 representatives).

1-62. **Detect.** Based on information from the decide portion of the targeting meeting, the G2/S2 should be prepared to do the following:

- Identify specific ISR assets.
- Develop IRs and propose pertinent IRs to the commander for designation as PIRs.
- Complete the initial ISR plan; ensure it addresses the HPTs.

1-63. **Electronic Warfare**. EW provides targeting information to the commander based on transmitter location data. There are three major subdivisions within EW: electronic support (ES), electronic protection (EP), and electronic attack (EA).

- **ES** – Also referred to as electronic warfare support (EWS) in JP 1-02, involves actions to search for, intercept, identify, and locate or localize sources of intentional and unintentional radiated electromagnetic energy. In addition to supporting the overall ISR effort, ES is needed to produce intelligence required to support EA missions.
- **EP** – Involves passive and active means taken to protect personnel, facilities, and equipment from any effects of friendly or enemy employment of EW that degrades, neutralizes, or destroys friendly combat capabilities.
- **EA** – Involves the use of electromagnetic energy, directed energy, or anti-radiation weapons to attack personnel, facilities, or equipment with the intent of degrading, neutralizing, or destroying enemy combat capability and is considered a form of fires. EA includes actions taken to prevent or reduce an enemy's effective use of the electromagnetic spectrum (EMS), such as jamming and electromagnetic deception.

1-64. **Deliver**. The G2/S2 should be prepared to do the following:

- Brief how the Intelligence warfighting function will work to detect and track the target through the entire targeting process.
- Brief the implementation of the intelligence synchronization and ISR plans, including the LTIOV and communications structure.

1-65. **Assess**. Depending on the factors of METT-TC and the PIRs, the same ISR assets that detect and track the targets may be required to support combat assessments in order to determine if re-attack is required. Conversely, if re-attack is a decision that the commander will make based upon the effects on the enemy, combat assessment should be reflected in the PIR, and supported in the corresponding ISR plan.

THE OPERATIONAL ENVIRONMENT

1-66. The operational environment can be defined as a composite of the conditions, circumstances, and influences that affect the employment of capabilities and bear on the decision of the commander. The operational environment encompasses physical areas and factors of the air, land, maritime, and space domains. It also includes the information environment and enemy, adversary, friendly, and neutral systems.

OPERATIONAL VARIABLES (PMESII-PT)

1-67. Analysis of the operational environment—in terms of political, military, economic, social, infrastructure, and information with the addition of physical environment and time (PMESII-PT) variables—provides relevant information that senior commanders use to understand, visualize, and describe the operational environment. As a set, these operational variables are often abbreviated as PMESII-PT. A description of each of the PMESII-PT variables follows:

- **Political.** Describes the distribution of responsibility and power at all levels of governance or cooperation.
- **Military.** Explores the military capabilities of all relevant actors in a given operational environment.
- **Economic.** Encompasses individual behaviors and aggregate phenomena related to the production, distribution, and consumption of resources.
- **Social.** Describes the cultural, religious, and ethnic makeup within an operational environment.
- **Information.** Describes the nature, scope, characteristics, and effects of individuals, organizations, and systems that collect, process, disseminate, or act on information.
- **Infrastructure.** Is composed of the basic facilities, services, and installations needed for the functioning of a community or society.
- **Physical Environment.** Defines the physical circumstances and conditions that influence the execution of operations throughout the domains of air, land, sea, and space.
- **Time.** Influences military operations within an operational environment in terms of the decision cycles, operational tempo, and planning horizons. (FM 3-0)

1-68. For the intelligence portion, when identifying threats based on the systems approach to PMESII-PT, there are four primary components of the operational environment. These are the threat, terrain, weather, and civil considerations.

1-69. While the operational variables are directly relevant to campaign planning, they are too broad to be applied directly to tactical planning. That does not mean they are not of value at all levels; they are fundamental to developing the understanding of the operational environment necessary to conduct planning at any level, in any situation. The degree to which each operational variable provides useful information depends on the situation and echelon. For example, social and economic variables often receive close analysis as part of enemy and civil considerations at brigade and higher levels. They may affect the training and preparation of small units. However, they are not relevant to a small-unit leader's mission analysis. That leader may only be concerned with such questions as "Who is the tribal leader for this village?" "Is the electrical generator working?" "Does the enemy have antitank missiles?"

MISSION VARIABLES

1-70. Upon receipt of a warning order (WARNO) or mission, Army leaders narrow their focus to six mission variables. Mission variables are those aspects of the operational environment that directly affect a mission. They outline the situation as it applies to a specific Army unit. The mission variables are mission, enemy, terrain and weather, troops and support available, time available, and civil considerations (METT-TC). These are the categories of relevant information used for mission analysis. Army leaders use the mission variables to synthesize operational variables and tactical-level information with local knowledge about conditions relevant to their mission. The METT-TC categories are listed below:

- **Mission.** The mission is the task, together with the purpose, that clearly indicates the action to be taken and the reason.
- **Enemy.** Relevant information regarding the enemy may include the following:
    - Threat characteristics (previously order of battle [OB] factors).
        - Doctrine (or known execution patterns).
        - Personal habits and idiosyncrasies.
        - Equipment, capabilities, and vulnerabilities.
    - Threat COAs.
- **Terrain and Weather.** Terrain and weather are natural conditions that profoundly influence operations. Terrain and weather are neutral; they favor neither side unless one is more familiar with—or better prepared to operate in—the environment.
- **Troops and Support Available.** The number, type, capabilities, and condition of available friendly troops and support available. These include the resources available from joint, interagency, intergovernmental, multinational, host nation (HN), commercial (via contracting), and private organizations.
- **Time Available.** Time is critical to all operations. Controlling and exploiting time is central to initiative, tempo, and momentum. By exploiting time, commanders can exert constant pressure, control the relative speed of decisions and actions, and force exhaustion on enemy forces.
- **Civil Considerations.** Civil considerations comprise six characteristics expressed in the memory aid ASCOPE:
    - Areas.
    - Structures.
    - Capabilities.
    - Organizations.
    - People.
    - Events.

Note. For additional information on ASCOPE and the IPB process, see FM 2-01.3. Understanding the operational environment requires understanding the civil aspects of the AO. Civil considerations reflect how the manmade infrastructure, civilian institutions, and the attitudes and activities of the civilian leaders, populations, and organizations within an AO influence the conduct of military operations.

1-71. METT-TC enables leaders to synthesize operational level information with local knowledge relevant to their missions and tasks in a specified AO. Tactical and operational leaders can then anticipate the consequences of their operations before and during execution. See FM 3-0 for a detailed discussion of PMESII-PT and METT-TC.

THE INTELLIGENCE PROCESS

1-72. Intelligence operations generally include the five functions that constitute the intelligence process: **plan, prepare, collect, process,** and

produce. Additionally, there are three common tasks that occur across the five functions of the intelligence process: **analyze, disseminate,** and **assess**. The three common tasks are discussed after the last function, produce. The intelligence process functions are not necessarily sequential; this is what differentiates the Army's intelligence process from the Joint intelligence cycle. The intelligence process provides a common model with which to guide one's thinking, discussing, planning, and assessing about the threat or AOI environment. The intelligence process generates information about the threat, the AOI, and the situation, which allows the commander and staff to develop a plan, seize and retain the initiative, build and maintain momentum, and exploit success.

PLAN

1-73. The plan step consists of the activities that identify pertinent IRs and develop the means for satisfying those requirements. The CCIRs (PIRs and FFIRs) drive the planning of the ISR effort. The intelligence officer supports the G3/S3 in arranging the ISR effort, based on staff planning, to achieve the desired collection effects. Planning activities include, but are not limited to—

- Conducting IPB.
- Managing requirements.
- Submitting RFIs and using intelligence reach to fill information gaps.
- Evaluating reported information.
- Establishing the intelligence communications and dissemination architecture.
- Developing, managing, and revising the intelligence synchronization plan and the ISR plan as mission requirements change.

PREPARE

1-74. The failure of MI units or personnel to accomplish their tasks or missions can often be attributed to their failure to prepare. The prepare step includes those staff and leader activities which take place upon receiving the operation order (OPORD), operation plan (OPLAN), warning order (WARNO), or commander's intent in order to improve the unit's ability to execute tasks or missions.

1-75. The most habitual and egregious preparation failures committed by leaders (as evidenced by performance at the combat training centers) is not the conduct of specific tasks, but the failure to adequately coordinate for more generic combat requirements such as—

- Friendly forward unit liaison.
- Departing or reentering friendly lines or AOs.
- Fire support: enroute, at mission location, and return.
- Casualty evacuation procedures.
- Passwords (running, forward of friendly lines), recognition signals, call signs, and frequencies.
- Resupply.
- Movement through friendly AOs.

COLLECT

1-76. The collect step involves collecting and reporting information in response to ISR tasking. ISR assets collect information and data about threat forces, activities, facilities, and resources as well as information concerning the environmental and geographical characteristics of a particular AO. A successful ISR effort results in the timely collection and reporting of relevant and accurate information. This collected information forms the foundation of intelligence databases, intelligence production, and the G2's/S2's situational understanding. The requirements manager evaluates the reported information for its responsiveness to the CCIRs (PIRs and FFIRs).

PROCESS

1-77. Processing involves converting collected data and information into a form that is suitable for analyzing and producing intelligence. Examples of processing include enhancing imagery, translating a document from a foreign language, converting electronic data into a standardized report that can be analyzed by a system operator, and correlating information.

1-78. Processing data and information is performed unilaterally and cooperatively by both humans and automated systems.

PRODUCE

1-79. The produce step involves evaluating, analyzing, interpreting, synthesizing, and combining information and intelligence from single or multiple sources into intelligence or intelligence products in support of known or anticipated requirements. Production also involves combining new information and intelligence with existing intelligence in order to produce intelligence in a form that the commander and staff can apply to the MDMP and that supports and helps facilitate situational understanding. During the produce step, the intelligence staff manipulates information by—

- Analyzing the information to isolate significant elements.
- Evaluating the information to determine accuracy, timeliness, usability, completeness, precision, and reliability.
- Combining the information with other relevant information and previously developed intelligence.
- Applying the information to estimate possible outcomes.
- Presenting the information in a format that will be most useful to its eventual user.

1-80. The intelligence staff deals with numerous and varied production requirements based on PIRs and intelligence requirements; diverse missions, environments, and situations; and presentation requirements. Through analysis, collaboration, and intelligence reach, the G2/S2 and the staff use the collective intelligence production capability of higher, lateral, and subordinate echelons to meet the production requirements. Proficiency in these techniques and procedures facilitates the intelligence staff's ability to answer command and staff requirements regardless of the factors of METT-TC.

Analyze

1-81. The three common tasks, discussed below, can occur throughout the intelligence process.

1-82. The intelligence staff analyzes intelligence and information about the enemy's capabilities, friendly vulnerabilities, and the battlefield environment as well as issues and problems that arise within the intelligence process itself to determine their nature, origin, and interrelationships. This analysis enables commanders, staffs, and leaders to determine the appropriate action or reaction and to focus or redirect assets and resources to fill information gaps or alleviate pitfalls. It is also within the analyze function that intelligence analysts sort through large amounts of collected information and intelligence to obtain only that information which pertains to the CCIRs (PIRs and FFIRs), maintenance of the COP, and facilitates the commander's situational understanding.

Disseminate

1-83. Disseminating is communicating relevant information of any kind from one person or place to another in a usable form by any means to improve understanding or to initiate or govern action. Disseminating intelligence entails using information management techniques and procedures to deliver timely, relevant, accurate, predictive, and usable intelligence to the commander. Determining the product format and selecting the means to deliver it are key aspects of dissemination. Information presentation may be in a verbal, written, interactive, or graphic format. The type of information, the time allocated, and the individual preference of the commander all influence the information format.

Assess

1-84. Assessment plays an integral role in all aspects of the intelligence process. Assessing includes evaluating the effectiveness of intelligence in supporting the operation. Assessing the situation and available information begins upon receipt of the mission and continues throughout the intelligence process. The continual assessment of intelligence operations and ISR efforts, available information and intelligence, and various aspects of the factors of METT-TC are critical to ensure—

- The G2/S2 answers all CCIRs (PIRs and FFIRs).
- The G2/S2 and G3/S3 redirect ISR assets to support changing requirements.
- Using information and intelligence properly.

INTELLIGENCE DISCIPLINES

1-85. Intelligence disciplines are categories of intelligence functions. The Army's intelligence disciplines are All-Source Intelligence, Human Intelligence (HUMINT), <u>Geospatial Intelligence (GEOINT),</u> Imagery Intelligence (IMINT), Signals Intelligence (SIGINT), Measurement and Signatures Intelligence (MASINT), <u>Open-Source Intelligence (OSINT),</u> Technical Intelligence (TECHINT), and Counterintelligence (CI). For more information regarding the intelligence disciplines, see Part Three of this manual, as well as the respective manuals, which covers each individual intelligence discipline.

ALL-SOURCE INTELLIGENCE

1-86. All-source intelligence is defined as the intelligence products, organizations, and activities that incorporate all sources of information and intelligence, including open-source information, in the production of intelligence. All-source intelligence is a separate intelligence discipline, as well as the name of the task used to produce intelligence from multiple intelligence or information sources.

HUMAN INTELLIGENCE

1-87. HUMINT is the collection of foreign information—by a trained HUMINT Collector—from people and multimedia to identify elements, intentions, composition, strength, dispositions, tactics, equipment, personnel, and capabilities. It uses human sources as a tool, and a variety of collection methods, both passively and actively, to collect information.

GEOSPATIAL INTELLIGENCE

1-88. Title 10 U.S. Code § 467 establishes GEOINT. The definition of geospatial intelligence is—

> The exploitation and analysis of imagery and geospatial information to describe, assess, and visually depict physical features and geographically referenced activities on the Earth. GEOINT consists of imagery, imagery intelligence, and geospatial information.

1-89. The Army implementation of GEOINT is a result of the Army's organization, manning, and training. There are multiple types of data and information that various Army units and organizations collect, provide, and analyze in order to support the GEOINT enterprise. The two primary GEOINT service providers in the Army are MI units and organizations and Engineer (topographic) units and organizations. MI units and organizations provide imagery and IMINT to the enterprise. Engineer (topographic) units and organizations provide geospatial data and information to the enterprise. Therefore, while some of the collection, analysis, and exploitation of imagery and geospatial information occur within the intelligence warfighting function; some of the collection, analysis, and exploitation of imagery and geospatial information occur outside intelligence.

IMAGERY INTELLIGENCE

1-90. IMINT is intelligence derived from the exploitation of imagery collected by visual photography, infrared, lasers, multi-spectral sensors, and radar. These sensors produce images of objects optically, electronically, or digitally on film, electronic display devices, or other media.

SIGNALS INTELLIGENCE

1-91. SIGINT is a category of intelligence comprising either individually or in combination all communications intelligence (COMINT), electronic intelligence (ELINT), and foreign instrumentation signals intelligence (FISINT), however transmitted. SIGINT is derived from communications, electronics, and foreign instrumentation signals.

MEASUREMENT AND SIGNATURES INTELLIGENCE

1-92. MASINT is technically derived intelligence that detects, locates, tracks, identifies, and/or describes the specific characteristics of fixed and dynamic target objects and sources. It also includes the additional advanced processing and exploitation of data derived from IMINT and SIGINT collection. MASINT collection systems include but are not limited to radar, spectroradiometric, electro-optical (E-O), acoustic, radio frequency (RF), nuclear detection, and seismic sensors, as well as techniques for gathering chemical, biological, radiological, and nuclear (CBRN), and other material samples.

OPEN-SOURCE INTELLIGENCE

1-93. The *National Defense Authorization Act for Fiscal Year 2006* states:

Open source intelligence is produced from publicly available information that is collected, exploited, and disseminated in a timely manner to an appropriate audience for the purpose of addressing a specific intelligence requirement.

1-94. Expressed in terms of the Army intelligence process, OSINT is relevant information derived from the systematic collection, processing, and analysis of publicly available information in response to intelligence requirements.

1-95. The Army does not have a specific military occupational specialty (MOS), additional skill identifier (ASI), or special qualification identifier (SQI) for OSINT. With the exception of the Asian Studies Detachment, the Army does not have base tables of organization and equipment (TOE) for OSINT units or staff elements. OSINT missions and tasks are imbedded within existing missions and force structure or accomplished through task organization.

TECHNICAL INTELLIGENCE

1-96. TECHINT is intelligence derived from the collection and analysis of threat and foreign military equipment and associated materiel for the purposes of preventing technological surprise, assessing foreign scientific and technical (S&T) capabilities, and developing countermeasures designed to neutralize an adversary's technological advantages.

COUNTERINTELLIGENCE

1-97. CI counters or neutralizes intelligence collection efforts through collection, counterintelligence investigations, operations, analysis, and production, and functional and technical services. CI includes all actions taken to detect, identify, track, exploit, and neutralize the multidiscipline intelligence activities of friends, competitors, opponents, adversaries, and enemies; and is the key intelligence community contributor to protect US interests and equities. CI assists in identifying EEFI, identifying vulnerabilities to threat collection, and actions taken to counter collection and operations against US forces.

UNIFIED ACTION

1-98. Unified action is the synchronization, coordination, and/or integration of the activities of governmental and nongovernmental entities with military operations to achieve unity of effort. (JP 1) Joint operations focus and maximize the complementary and reinforcing effects and capabilities of each service. Joint force commanders (JFCs) synchronize the complementary capabilities of the service components that comprise the joint force.

1-99. The employment of MI in campaigns and major operations must be viewed from a joint perspective, and the "mud-to-space" intelligence concept must establish a fully interoperable and integrated joint intelligence capability. ARFOR intelligence assets work with multinational and interagency partners to accomplish their missions. Ideally, multinational and interagency intelligence partners provide cultures, perspectives, and capabilities that reinforce and complement Army MI strengths and capabilities. Close intelligence coordination is the foundation of successful unified action. (See chapter 2.)

1-100. A critical part of current operations is the execution of the joint doctrinal concept of persistent surveillance. Joint doctrine defines persistent surveillance as a collection strategy that emphasizes the ability of some collection systems to linger on demand in an area to detect, locate, characterize, identify, track, target, and possibly provide battle damage assessment and re-targeting in real or near-real time. Persistent surveillance facilitates the formulation and execution of preemptive activities to deter or forestall anticipated adversary courses of action (JP 2-0).

THE NATURE OF LAND OPERATIONS

1-101. Landpower is the ability—by threat, force, or occupation—to gain, sustain, and exploit control over land, resources, and people (FM 3-0). Army operations reflect expeditionary and campaign capabilities that constantly adapt to each campaign's unique circumstances. Expeditionary capabilities require forces organized to be modular, versatile, and rapidly deployable. Rapidly deployed expeditionary force packages provide immediate options for seizing or retaining the operational initiative and also allow the conduct of sustained operations for as long as necessary. Army forces are organized, trained, and equipped for endurance. The Army's preeminent challenge is to balance expeditionary agility and responsiveness with the endurance and adaptability needed to complete a campaign successfully.

1-102. The capability to prevail in close combat is indispensable and unique to land operations. The outcome of battles and engagements depends on Army forces' ability to prevail in close combat. Many factors inherent in land combat combine to complicate the situation. These factors include chaos, complexity, insufficient intelligence, errors in understanding or planning, difficult terrain, the civilian population, and an adaptive and lethal enemy. The axiom "intelligence drives operations" continues to be true especially for land operations; operations and intelligence are inextricably linked.

1-103. Four considerations are preeminent in generating expeditionary capabilities and Army force packages.

- **Scope.** Considers and strives to understand an enemy throughout the depth of an operational area. Commanders rely on intelligence in order to use maneuver, fires, and other elements of combat power to defeat or destroy enemy forces.
- **Duration.** Forces routinely conduct missions prior to, during, and after the commitment of land combat forces. Intelligence is always engaged.
- **Terrain.** Missions occur among a complex variety of natural and manmade features. Employing forces in the complexity of the ground environment requires thorough planning.
- **Permanence.** Forces are integrated with, or assigned to, land combat forces as they seize or secure ground.

1-104. Several attributes of the land environment affect the application of landpower. These attributes include—

- The requirement to deploy and employ Army forces rapidly.
- The requirement for Army forces to operate for protracted periods.
- The nature of close combat.
- Uncertainty, chance, friction, and complexity.

1-105. Reconnaissance, Surveillance, and Target Acquisition (RSTA)/ISR is the means the Army uses to implement the joint doctrinal concept of persistent surveillance in support of tactical operations. Dependable technology and responsive intelligence lessen the effects of uncertainty, chance, friction, and complexity. Complex and dynamic Army tactical operations require extensive ISR capabilities to satisfy the commander's information requirements to detect, locate, characterize, identify, track, and target HPTs, and to provide combat assessment in real time within a very fluid operational environment.

1-106. *Reconnaissance, Surveillance, and Target Acquisition/ Intelligence, Surveillance, and Reconnaissance* **is a full spectrum combined arms mission that integrates ground and air capabilities to provide effective, dynamic, timely, accurate, and assured combat information and multidiscipline actionable intelligence for lethal and non-lethal effects and decisions in direct support of the ground tactical commander.**

FORCE PROJECTION OPERATIONS

1-107. Force projection is the military component of power projection. It is a central element of the national military strategy. Army organizations and installations, linked with joint forces and industry, form a strategic platform to maintain, project, and sustain ARFOR, wherever they deploy. Force projection operations are inherently joint and require detailed planning and synchronization. Force projection encompasses a range of processes—mobilization, deployment, employment, sustainment, and redeployment—discussed below.

1-108. The Army must change its mindset from depending on an "intelligence buildup" to performing intelligence readiness on a daily basis in order to meet the requirements for strategic responsiveness. MI personnel, even in garrison at the lowest tactical echelons, must use their analytic and other systems and prepare for possible operations on a daily basis.

1-109. Built on a foundation of intelligence readiness, the Intelligence warfighting function provides the commander with the intelligence needed to plan, prepare, and execute force projection operations. Successful intelligence during force projection operations relies on continuous collection and intelligence production before and during the operation. In a force projection operation, higher echelons will provide intelligence to lower echelons until the tactical ground force completes entry and secures the lodgment area. The joint force J2 must exercise judgment when providing information to subordinate G2s/S2s so not to overwhelm them. Figure 1-5 is an example of force projection intelligence.

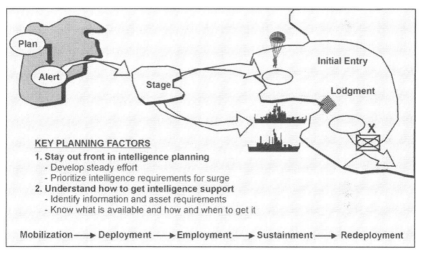

Figure 1-5. Force Projection Intelligence.

1-110. The G2/S2 must anticipate, identify, consider, and evaluate all threats to the entire unit throughout force projection operations. This is critical during the deployment and entry operations stages of force projection. During these stages, the unit is particularly vulnerable to enemy actions because of its limited combat power and knowledge of the AO. Intelligence personnel must, therefore, emphasize the delivery of combat information and intelligence products that indicate changes to the threat or AO developed during predeployment IPB.

1-111. The G2/S2 should—

FM 2-0, C1

- Review available databases on assigned contingency AOIs, conduct IPB on these AOIs, and develop appropriate IPB products.
- Comply with higher headquarters standing operating procedures (SOPs) and manuals for specific intelligence operations guidance.
- Coordinate for and rehearse electronic message transfers (for example, Internet Protocol addresses, routing indicators) using the same communications protocols with theater, higher headquarters, subordinate, and lateral units that the unit would use when deployed.
- Plan, train, and practice surging intelligence functions on likely or developing contingency crises.
- Prepare and practice coordination from predeployment through redeployment with other elements and organizations (for example, HUMINT, IMINT, SIGINT, MASINT, IO, staff weather officer [SWO], civil affairs [CA], psychological operations [PSYOP], and special operations forces [SOF] units, to include databases and connectivity).
- Include the following as a part of daily (sustainment) operations:
  - US Army Reserve and other augmentation.
  - A linguist plan with proficiency requirements. (Alert linguists through early entry phases of deployment.)
  - Training (individual and collective).
- Establish formal or informal intelligence links, relationships, and networks to meet developing contingencies.
- Forward all RFIs to higher headquarters in accordance with SOPs.
- Establish statements of intelligence interests (SIIs), other production, and I&W requirements.

1-112. To draw intelligence from higher echelons and focus intelligence downward, based on the commander's needs, the G2/S2 must—

- Understand the J2's multiple echelon and broadcast dissemination capability to ensure near-real time (NRT) reporting to all deployed, in transit, or preparing to deploy forces.
- Maintain or build intelligence databases on the environment and threats for each probable contingency.
- State and record the CCIR (as a minimum, list the PIRs and ISR tasks or requests).

1-113. Until the unit's collection assets become operational in the AO, the G2/S2 will depend upon intelligence from the ARFOR or JTF to answer the unit's intelligence needs. The following paragraphs describe the intelligence and ISR considerations during force projection.

MOBILIZATION

1-114. Mobilization is the process by which the armed forces or part of them are brought to a state of readiness for war or other national emergency. It assembles and organizes resources to support national objectives. Mobilization includes activating all or part of the US Army Reserve, and assembling and organizing personnel, supplies, and materiel. A unit may be brought to a state of readiness for a specific mission or other national emergency. This process, called mobilization, is where specific US Army

Active or Reserve units, capabilities, and personnel are identified and integrated into the unit. During mobilization, the G2/S2 must—
- Monitor intelligence reporting on threat activity and I&W data.
- Manage IRs and RFIs from their unit and subordinate units to include updating ISR planning.
- Establish habitual training relationships with their US Army Active and Reserve augmentation units and personnel as well as higher echelon intelligence organizations as identified in the existing OPLAN. Support the US Army Reserve units and augmentation personnel by preparing and conducting intelligence training and threat update briefings and by disseminating intelligence.
- Identify ISR force requirements for the different types of operations and contingency plans (CONPLANs).
- Identify individual military, civilian, and contractor augmentation requirements for intelligence operations. The Army, and the Intelligence warfighting function in particular, cannot perform its missions without the support of its Department of the Army Civilians (DACs) and contractors. The force increasingly relies on the experience, expertise, and performance of non-uniformed personnel and has fully integrated these non-uniformed personnel into the warfighting team.

1-115. During mobilization the G2/S2, in conjunction with the rest of the staff, must ensure the adequacy of training and equipping of US Army Active and Reserve MI organizations and individual augmentees to conduct intelligence operations.

1-116. The G2/S2 supports peacetime contingency planning with IPB products and databases on likely contingency areas. The G2/S2 establishes an intelligence synchronization plan that will activate upon alert notification. For smooth transition from predeployment to entry, the G2/S2 must coordinate intelligence synchronization and communications plans before the crisis occurs. The intelligence synchronization plan identifies the intelligence requirements supporting those plans, to include—
- ISR assets providing support throughout the AOI.
- Command and support relationships of ISR assets at each echelon.
- Report and request procedures not covered in unit SOPs.
- Sequence of deployment of ISR personnel and equipment. Early deployment of key ISR personnel and equipment is essential for force protection and combat readiness. Composition of initial and follow-on deploying assets is influenced by the factors of METT-TC, availability of communications, and availability of lift.
- Communications architecture supporting both intelligence staffs and ISR assets.
- Friendly vulnerabilities to hostile intelligence threats and plans for conducting FP measures. The staff must begin this type of planning as early as possible to ensure adequate support to FP of deploying and initial entry forces.
- Monitor time-phased force and deployment data (TPFDD) and recommend changes in priority of movement, unit, or capability to enable ISR operations.

1-117. The G2/S2 must continually monitor and update the OPLANs to reflect the evolving situation, especially during crisis situations. National intelligence activities monitor regional threats throughout the world and can answer some intelligence requirements supporting the development of OPLANs.

1-118. Upon alert notification, the G2/S2 updates estimates, databases, IPB products, and other intelligence products needed to support command decisions on force composition, deployment priorities and sequence, and the AOI. Units reassess their collection requirements immediately after alert notification. The G2/S2 begins verifying planning assumptions within the OPLANs. CI and ISR personnel provide FP support and antiterrorism measures.

1-119. Throughout mobilization, unit intelligence activities will provide the deploying forces with the most recent intelligence on the AO. The intelligence staff will also update databases and situation graphics. The G2/S2 must—

- Fully understand the unit, ARFOR, and joint force intelligence organizations.
- Revise intelligence and intelligence-related communications architecture and delete or integrate any new systems and software with the current architecture.
- Support 24-hour operations and provide continuous intelligence.
- Plan all required intelligence reach procedures.
- Determine transportation availability for deployment and availability when deployed.
- Determine all sustainability requirements.
- Determine intelligence release requirements and restrictions; releasability to multinational and HN sources.
- Review status of forces agreements (SOFAs), rules of engagement (ROE), international laws, and other agreements, emphasizing the effect that they have on intelligence collection operations. (Coordinate with the staff judge advocate [SJA] on these issues.)
- Ensure ISR force deployment priorities are reflected in the TPFDD to support ISR operations based upon the factors of METT-TC.
- Ensure intelligence links provide the early entry commander vital access to multi-source army and joint intelligence collection assets, processing systems, and databases.
- Review the supporting unit commanders' specified tasks, implied tasks, task organization, scheme of support, and coordination requirements with forward maneuver units. Address issues or shortfalls and direct or coordinate changes.
- Establish access to national HUMINT, IMINT, SIGINT, MASINT, and CI databases, as well as automated links to joint service, multinational, and HN sources.

DEPLOYMENT

1-120. Deployment is the movement of forces and materiel from their point of origin to the AO. This process has four supporting components:

predeployment activities, fort to port, port to port, and port to destination; these components are known collectively as reception, staging, onward movement, and integration (RSO&I) activities. Success in force projection operations hinges on timely deployment. The size and composition of forces requiring lift are based on the factors of METT-TC, availability of pre-positioned assets, the capabilities of HN support, and the forward presence of US forces. Force or tactical tailoring is the process used to determine the correct mix and sequence of deploying units.

1-121. During deployment, intelligence organizations at home station or in the rear area take advantage of modern SATCOM, broadcast technology, and automated data processing (ADP) systems to provide graphic and textual intelligence updates to the forces enroute. Enroute updates help eliminate information voids and, if appropriate, allow the commander to adjust the plan prior to arrival in theater in response to changes in the OE or enemy actions.

1-122. Intelligence units extend established networks to connect intelligence staffs and collection assets at various stages of the deployment flow. Where necessary, units establish new communications paths to meet unique demands of the mission. The theater and corps analysis and control elements (ACEs) play a critical role in making communications paths, networks, and intelligence databases available to deploying forces.

1-123. Space-based systems are key to supporting intelligence during the deployment and the subsequent stages of force projection operations by—

- Monitoring terrestrial AOIs through ISR assets to help reveal enemy location and disposition, attempting to identify the enemy's intent.
- Providing communications links between forces enroute and in the continental United States (CONUS).
- Permitting MI collection assets to accurately determine their position through the Global Positioning System (GPS).
- Providing timely and accurate data on meteorological, oceanographic, and space environmental factors that might affect operations.
- Providing warning of theater ballistic missile launches.
- Providing timely and accurate weather information to all commanders through the Integrated Meteorological System (IMETS).

1-124. Situation development dominates intelligence operations activities during initial entry operations. The G2/S2 attempts to identify all threats to arriving forces and assists the commander in developing FP measures. During entry operations, echelons above corps (EAC) organizations provide intelligence. This support includes providing access to departmental and joint intelligence and deploying scalable EAC intelligence assets. The entire effort focuses downwardly to provide tailored support to deploying and deployed echelons in response to their CCIRs (PIRs and FFIR).

1-125. Collection and processing capabilities are enhanced, as collection assets build up in the deployment area, with emphasis on the build-up of the in-theater capability required to conduct sustained ISR operations. As the build-up continues, the G2/S2 strives to reduce total dependence on extended split-based intelligence from outside the AO. As assigned collection assets arrive into the theater, the G2/S2 begins to rely on them for tactical intelligence although higher organizations remain a source of intelligence.

1-126. As the ARFOR enter the theater of operations, the joint force J2 implements and, where necessary, modifies the theater intelligence architecture. Deploying intelligence assets establishes liaison with staffs and units already present in the AO. Liaison personnel and basic communications should be in place prior to the scheduled arrival of parent commands. ISR units establish intelligence communications networks.

1-127. CONUS and other relatively secure intelligence bases outside the AO continue to support deployed units. Systems capable of rapid receipt and processing of intelligence from national systems and high capacity, long-haul communications systems are critical to the success of split-based support of a force projection operation. These systems provide a continuous flow of intelligence to satisfy many operational needs.

1-128. The G2/S2, in coordination with the G3/S3, participates in planning to create conditions for decisive operations. The G2/S2 also adjusts collection activities as combat strength builds. During entry operations the G2/S2—

- Monitors FP indicators.
- Monitors the ISR capability required to conduct sustained intelligence operations.
- Monitors intelligence reporting on threat activity and I&W data.
- Develops measurable criteria to evaluate the results of the intelligence synchronization plan.
- Assesses—
  - Push versus pull requirements of intelligence reach.
  - Effectiveness of the intelligence communications architecture.
  - Reporting procedures and timelines.
  - Intelligence to OPLANs and OPORDs, branches, and sequels (to include planning follow-on forces).

EMPLOYMENT

1-129. Employment is the conduct of operations to support a JFC commander. Employment encompasses an array of operations, including but not limited to—

- Entry operations (opposed or unopposed).
- Shaping operations (lethal and non-lethal).
- Decisive operations (combat or support).
- Postconflict operations (prepare for follow-on missions or redeployment).

Entry Operations

1-130. Enemies often possess the motives and means to interrupt the deployment flow of ARFOR. Threats to deploying forces may include advanced conventional weaponry (air defense, mines, etc.) and WMD. Sea and air ports of debarkation (PODs) should be regarded as enemy HPTs because they are the entry points for forces and equipment. PODs are vulnerable because they are fixed targets with significant machinery and

FM 2-0, C1

equipment that is vulnerable to attack; in addition to military forces and materiel, HN support personnel, contractors, and civilians may all be working there. An enemy attack, or even the threat of an enemy attack, on a POD can have a major impact on force projection momentum. Commanders at all levels require predictive intelligence so that they may focus attention on security actions that reduce vulnerabilities. To avoid, neutralize, or counter threats to entry operations, the commanders rely on the ability of the G2/S2 to support future operations by accurately identifying enemy reactions to US actions, anticipating their response to our counteractions and predicting additional ECOAs.

1-131. Predictive intelligence also supports the decisions the commander and staff must make about the size, composition, structure, and deployment sequence of the force in order to create the conditions for success. Commanders rely on predictive intelligence to identify potential friendly decisions before the actual event. While thorough planning develops friendly COAs to meet possible situations, the nature of an operation can change significantly before execution. The G2/S2 must provide timely, accurate, and predictive intelligence to ensure the commander can retain the initiative to implement the plan or make decisions before losing the opportunity to do so.

Shaping Operations

1-132. Shaping operations create and preserve conditions for the success of the decisive operation. Shaping operations include lethal and non-lethal activities conducted throughout the AO. They support the decisive operation by affecting enemy capabilities and forces, or by influencing enemy decisions. G2/S2 intelligence analysis and ISR activities support the development and execution of shaping operations by identifying threat centers of gravity and decisive points on the battlefield. The G2/S2 also ensures the intelligence process focuses on the CCIRs (PIRs and FFIRs). It is critical that the G2/S2 provide timely, accurate, and predictive intelligence to the commander to facilitate the execution of shaping operations. Predictive intelligence should provide sufficient time for the commander to understand how the enemy will react to US COAs so that appropriate shaping operations can be implemented.

Decisive Operations

1-133. Decisive operations are those that directly accomplish the task assigned by the higher headquarters. Decisive operations conclusively determine the outcome of major operations, battles, and engagements. Continuously synchronized ISR activities, coupled with predictive intelligence results and products during all stages of force projection, combine to ensure the commander is prepared to employ the right forces with the right support to conduct decisive operations at the most appropriate place and time. In addition to coordinating with the G3/S3 to synchronize all ISR activities, the G2/S2 must answer the CCIRs (PIRs and FFIRs) and provide continuous assessments of the enemy's current situation and predict the enemy's subsequent COA, branches, and/or sequels.

Postconflict Operations

1-134. Upon cessation of hostilities or truce, deployed forces enter a new stage of force projection operations. Postconflict operations focus on restoring order, reestablishing HN infrastructure, preparing for redeployment of forces, and planning residual presence of US forces. While postconflict operations strive to transition from conflict to peace, there remains a possibility of resurgent hostilities by individuals and forces. The ISR effort, particularly predictive intelligence analysis, remains just as critical in postconflict operations as in the other employment operations. ISR operations support the postconflict emphasis on restoration operations exemplified by commanders redirecting their CCIRs (PIRs and FFIRs) and IRs to support units conducting these missions. These operations might include—

- Engineer units conducting mine clearing or infrastructure reconstruction operations.
- Medical and logistics units providing humanitarian relief.
- MP units providing law and order assistance.
- CA units reestablishing local control and preparing for the orderly transition of local governments and to support civil-military operations (CMO).

SUSTAINMENT

1-135. Sustainment involves providing and maintaining levels of personnel and materiel required to sustain the operation throughout its duration. It is essential to generating combat power. CSS may be split-based between locations within and outside CONUS. These operations include ensuring units have the MI assets required to accomplish the mission, such as personnel (including linguists), communications systems, ISR systems, and appropriate maintenance support.

REDEPLOYMENT

1-136. Redeployment is the process by which units and materiel reposture themselves in the same theater; transfer forces and materiel to support another JFC's operational requirements; or return personnel and materiel to the home or demobilization station upon completion of the mission. Redeployment operations encompass four phases:

- Recovery, reconstitution, and pre-deployment activities.
- Movement to and activities at the port of embarkation (POE).
- Movement to the POD.
- Movement to home station.

1-137. As combat power and resources decrease in the AO, FP and I&W become the focus of the commander's intelligence requirements. This in turn drives the selection of those assets that must remain deployed until the end of the operation and those that may redeploy earlier. The S2—

- Monitors intelligence reporting on threat activity and I&W data.
- Continues to conduct intelligence to FP.
- Requests ISR support (theater and national systems) and intelligence in support of redeployment.

1-138. After redeployment, MI personnel and units recover and return to predeployment activities. ISR units resume contingency-oriented peacetime intelligence operations. US Army Reserve ISR units demobilize and return to peacetime activities. G2/S2s must—
- Monitor intelligence reporting on threat activity and I&W data.
- Update or consolidate databases.
- Maintain intelligence readiness.
- Provide their input into the Force Design Update (FDU) process to refine modified table of organizations and equipment (MTOE) and evaluate the need for Individual Mobilization Augmentee (IMA) personnel.

Chapter 2

Intelligence and Unified Action

UNIFIED ACTION

2-1. As discussed in JP 1-02 and Chapter 1, unified action describes the wide scope of actions (including the synchronization of activities with governmental and NGO agencies) taking place within unified commands, subordinate unified commands, or JTFs under the overall direction of the commanders of those commands. Under unified action, commanders integrate joint, single-service, special, and supporting intelligence operations with interagency, nongovernmental, and multinational operations. The ARFOR often brings unique ISR capabilities to unified action operations.

2-2. In a unified action, the intelligence staff takes on additional responsibilities, relationships, and infrastructure to execute multi-service and multinational intelligence operations. This chapter discusses the synchronization of Army intelligence efforts with joint and other national and international partners to achieve unity of effort and to accomplish the commander's intent.

THE LEVELS OF WAR

2-3. The levels of war are doctrinal perspectives that clarify the links between strategic objectives and tactical actions. Although there are no finite limits or boundaries between them, the three levels of war are strategic, operational, and tactical (see Figure 2-1).

- The strategic level is that level at which a nation, often as one of a group of nations, determines national and multinational security objectives and guidance and develops and uses national resources to accomplish them.
- The operational level is the level at which campaigns and major operations are conducted and sustained to accomplish strategic objectives within theaters or AOs. It links the tactical employment of forces to strategic objectives.
- The tactical level is the employment of units in combat. It includes the ordered arrangement and maneuver of units in relation to each other, the terrain, and the enemy to translate potential combat power into victorious battles and engagements.

2-4. Understanding the interdependent relationship of all three levels of war helps commanders visualize a logical flow of operations, allocate resources, and assign tasks. Actions within the three levels are not associated with a particular command level, unit size, equipment type, or force or component type. The concept of strategic, operational, and tactical intelligence operations aids JFCs and their J2s in visualizing the flow of intelligence from one level to the next. The concept facilitates allocating required collection, analysis, production, and dissemination resources; and facilitates the

FM 2-0

assignment of appropriate intelligence tasks to national, theater, component, and supporting intelligence elements.

STRATEGIC

2-5. The President and the Secretary of Defense use strategic intelligence to develop national strategy and policy, monitor the international situation, prepare military plans, determine major weapon systems and force structure requirements, and conduct strategic operations.

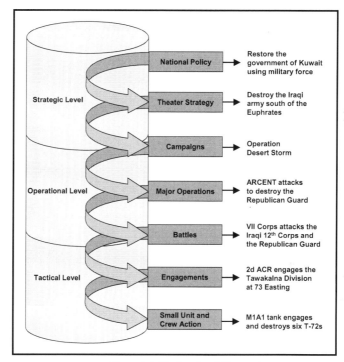

Figure 2-1. The Levels of War.

2-6. Intelligence supports joint operations across the full spectrum of military operations. It determines the current capabilities, and forecasts future developments, of adversaries or potential adversaries that could affect the national security and interests. A strategic operation also produces the intelligence required by combatant commanders to prepare strategic estimates, strategies, and plans to accomplish missions assigned by higher authorities. Theater intelligence includes determining when, where, and in what strength the adversary will stage and conduct theater level campaigns

and strategic unified operations. The intelligence staff should also focus predictive analysis efforts on identifying strategic threat events and how these events will impact US actions at the strategic, operational, and tactical levels. Intelligence operations support IO as well. Intelligence operations support strategic planning by—
- Developing strategic intelligence policy.
- Preparing the strategic collection plan.
- Allocating national intelligence resources.

2-7. Intelligence requirements are those requirements generated from the staff's IRs regarding the enemy and environment that are not a part of the CCIRs (PIRs and FFIRs). Intelligence requirements require collection and can provide answers in order to identify indicators of enemy actions or intent, which reduce the uncertainties associated with an operation. Significant changes (that is, branches and sequels) with an operation usually lead to changes in intelligence requirements. Of particular importance is information relating to enemy or threat strategic vulnerabilities, strategic forces, strategic centers of gravity, and any capabilities relating to the development and employment of CBRNE.

2-8. Global and regional issues and threats are identified and reported to the President and the Secretary of Defense, as well as to the senior military leadership and the combatant commanders. Intelligence requirements include any foreign developments that could threaten the US, its citizens abroad, or multinational military, political, or economic interests. Intelligence also includes identifying hostile reactions to US reconnaissance activities and indications of impending terrorist attacks (I&W). For a complete listing of strategic tasks, refer to CJCSM 3500.04C.

OPERATIONAL

2-9. Combatant commanders and subordinate JFCs and their component commanders are the primary users of intelligence. At the operational echelons, intelligence—
- Focuses on the military capabilities and intentions of enemies and threats.
- Provides analysis of events within the AOI and helps commanders determine when, where, and in what strength the adversary might stage and conduct campaigns and major operations.
- Supports all phases of military operations, from mobilization all the way through redeployment of US forces, and continues during sustainment.
- Supports all aspects of the joint campaign.
- Identifies adversary centers of gravity and HVTs.
- Provides critical support to friendly IO.

2-10. The JFC and staff allocate intelligence resources and request support from national agencies, other theaters, and multinational partners. During stability operations and support operations, operational intelligence includes training and assisting multinational partners in conducting intelligence operations.

2-11. Combatant commanders use intelligence concerning the nature and characteristics of the battlespace to determine the type and scale of operations. Intelligence also aids in determining the impact of significant regional features and hazards on the conduct of both friendly and adversary operations. Significant regional factors include the natural environment, political, informational, economic, industrial, infrastructure, geographic, demographic, topographic, hydrographic, climatic, populace, cultural, medical, lingual, historical, and psychological aspects of the AOI. Intelligence analysis also assists in determining the ROE and other restrictions which will affect operations in the JFC's AO.

2-12. Intelligence relating to the adversary's military and nonmilitary capabilities assists in determining the adversary's ability to conduct military operations. Factors that operational intelligence addresses include mobilization potential, force structure (including alliance forces), force dispositions, equipment, military doctrine, C2 structure, and their MDMP. Intelligence includes the continuous refinement of the OBs for the entire array of the enemy's forces in the AOI. For a complete listing of operational tasks, refer to CJCSM 3500.04C.

US Army Intelligence and Security Command (INSCOM)

2-13. The U.S. Army has vested its intelligence at the operational level with INSCOM, the major Army command (MACOM) responsible for the Army's intelligence forces above Corps. INSCOM's mission is to conduct and support dominant intelligence, security, and information operations for military commanders and national decisionmakers. The INSCOM strategy is to provide superior information and information capabilities to Army commanders, while denying the same to adversaries. Headquarters (HQ), INSCOM, in coordination with its major subordinate commands (MSCs), provides a myriad of general intelligence support operations. INSCOM is providing a globally focused, rapidly deployable, knowledge based, adaptively force packaged capability, supporting commanders and leaders with actionable intelligence at the point of decision. INSCOM serves as the national to tactical Intelligence bridge.

Army Space Program Office (ASPO)

2-14. ASPO executes the Army's Tactical Exploitation of National Capabilities Program (TENCAP). The program focuses on exploiting current and future tactical potential of national systems and integrating the capabilities into the Army's tactical decisionmaking process. Army TENCAP systems enable the tactical commander maximum flexibility to satisfy intelligence needs under a wide range of operational scenarios. ASPO is the point of contact (POC) for all tactical activities between MACOMs or users and the National Reconnaissance Office (NRO).

TACTICAL

2-15. Tactical commanders use intelligence for planning and conducting battles and engagements. Relevant, accurate, predictive, and timely intelligence allows tactical units to achieve an advantage over their adversaries. Precise and predictive intelligence, on the threat and targets, is essential for mission success. Predictive intelligence also enables the staff to better identify or develop ECOAs. Tactical intelligence—

- Identifies and assesses the enemy's capabilities, COAs, and vulnerabilities, as well as describes the battlespace.
- Seeks to identify when, where, and in what strength the enemy will conduct tactical level operations.
- Provides the commander with information on imminent threats to the force including those from terrorists, saboteurs, insurgents, and foreign intelligence collection.
- Provides critical support to friendly IO.
- Develops and disseminates targeting information and intelligence.

2-16. Intelligence provides the tactical commander with the information and intelligence he requires to successfully employ combat forces against enemy forces. Thus, intelligence tasks support the execution of battles and engagements. These intelligence tasks are different from those at other levels due to their ability to immediately influence the outcome of the tactical commander's mission. They include information gathered from tactical sources, such as combat information, interrogations, debriefings, and eliciting information from captured or misplaced personnel. For a complete listing of tactical tasks, refer to FM 7-15.

CATEGORIES OF INTELLIGENCE

2-17. Directly or indirectly, the unit receives intelligence from throughout the US intelligence community. The intelligence operations and products of national, joint, and service organizations that make up the intelligence community fall into one of six categories:

- I&W.
- Current intelligence.
- GMI.
- Target intelligence.
- S&T intelligence.
- CI.

2-18. The categories of intelligence are distinguishable from each other primarily by the purpose of the intelligence product. The categories can overlap and some of the same intelligence is useful in more than one category. Depending upon the echelon, intelligence organizations use specialized procedures to develop each category of intelligence. The following information describes each category.

INDICATIONS AND WARNINGS

2-19. Indications and warnings (I&W) are those intelligence activities intended to detect and report time-sensitive intelligence information on foreign developments that could involve a threat to the US or allied and/or coalition military, political, or economic interests or to US citizens abroad. I&W includes forewarning of enemy actions or intentions; the imminence of hostilities; insurgency; nuclear or non-nuclear attack on the US, its overseas forces, or allied and/or coalition nations; hostile reactions to US reconnaissance activities; terrorist attacks; and other similar events. (JP 3-13). I&W comes from time-sensitive information and analysis of developments that could involve a threat to the US and multinational military forces, US political or economic interests, or to US citizens. While the G2/S2 is primarily responsible for producing I&W intelligence, each element, such as the MPs conducting PIO, within every unit contributes to I&W through awareness of the CCIRs and reporting related information.

CURRENT INTELLIGENCE

2-20. Current intelligence involves the integration of time-sensitive, all-source intelligence and information into concise, accurate, and objective reporting on the battlespace and current enemy situation. One of the most important forms of current intelligence is the enemy situation portion of the COP. The G2/S2 is responsible for producing current intelligence for the unit. Current intelligence supports ongoing operations across full spectrum operations. In addition to the current situation, current intelligence should provide projections of the enemy's anticipated situations (estimates) and their implications on the friendly operation.

GENERAL MILITARY INTELLIGENCE

2-21. GMI is intelligence concerning military capabilities of foreign countries or organizations or topics affecting potential US or multinational military operations relating to armed forces capabilities, including OB, organization, training, tactics, doctrine, strategy, and other factors bearing on military strength and effectiveness and area and terrain intelligence. This broad category of intelligence is normally associated with long-term planning at the national level. However, GMI is also an essential tool for the intelligence staff and should be in place long before the start of preparations for a particular military operation. An up-to-date, comprehensive intelligence database is critical to the unit's ability to plan and prepare rapidly for the range of operations and global environments in which it may operate. GMI supports the requirement to quickly respond to differing crisis situations with corresponding intelligence spanning the globe. One of the many places to get information for GMI is the medical intelligence database. The G2/S2 planner develops his initial IPB from GMI products.

2-22. The G2/S2 develops and maintains the unit's GMI database on potential threat forces and environments based on the commander's guidance. As an essential component of intelligence readiness, this database supports the unit's planning, preparation, execution, and assessment of operations. The G2/S2 applies and updates the database as it executes its intelligence production tasks.

TARGET INTELLIGENCE

2-23. Target intelligence is the analysis of enemy units, dispositions, facilities, and systems to identify and nominate specific assets or vulnerabilities for attack, re-attack, or exploitation (for intelligence). It consists of two mutually supporting production tasks: target development and combat assessment.

- **Target development** is the systematic evaluation and analysis of target systems, system components, and component elements to determine HVTs for potential attack through maneuver, fires, or non-lethal means.
- Once attacked, **combat assessment** provides a timely and accurate estimate of the affects of the application of military force (lethal or non-lethal) an IO on targets and target systems based on predetermined objectives.

SCIENTIFIC AND TECHNICAL INTELLIGENCE

2-24. Scientific and technical intelligence (S&TI) is the product resulting from the collection, evaluation, analysis, and interpretation of foreign S&T information which covers foreign developments in basic and applied research and in applied engineering techniques and S&T characteristics, capabilities, and limitations of all foreign military systems, weapons, weapon systems, and materiel, the research and development (R&D) related thereto, and the production methods employed for their manufacture.

2-25. S&T intelligence concerns foreign developments in basic and applied sciences and technologies with warfare potential. It includes characteristics, capabilities, vulnerabilities, and limitations of all weapon systems, subsystems, and associated materiel, as well as related R&D. S&T intelligence also addresses overall weapon systems and equipment effectiveness. Specialized organizations—such as the Defense Intelligence Agency (DIA) Missile and Space Intelligence Center (MSIC), INSCOM, Air Missile Defense (AMD), Area Air and Missile Defense Command (AAMDC), and National Ground Intelligence Center (NGIC)—produce this category of intelligence. The G2/S2 establishes instructions within SOPs, orders, and plans for handling and evacuating captured enemy materiel (CEM) for S&T intelligence (S&TI) exploitation.

COUNTERINTELLIGENCE

2-26. As previously defined at the end of Chapter 1, CI attempts to identify and recommend countermeasures to the threat posed by foreign intelligence security services (FISS) and the ISR activities of non-state entities such as organized crime, terrorist groups, and drug traffickers. CI analysis incorporates information from all sources as well as the results of CI collection, investigations, and operations to analyze the multidiscipline threat posed by foreign intelligence services and activities.

INTELLIGENCE COMMUNITY

2-27. There are many organizations in the intelligence community that support military operations by providing specific intelligence products and services. The J2/G2/S2 and his staff must be familiar with these organizations and the methods of obtaining information from them as necessary. Figure 2-2 shows organizations that compose the intelligence community.

FM 2-0

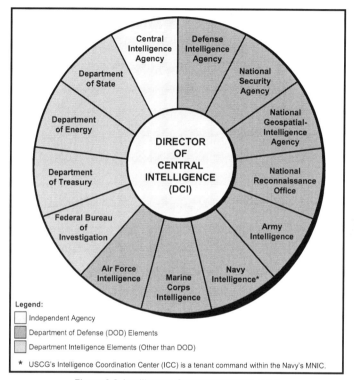

Figure 2-2. Intelligence Community Membership.

DOD AGENCIES

Defense Intelligence Agency (DIA)

2-28. DIA is a combat support agency and a major collector and producer in the defense intelligence community. The DIA supports the full spectrum of operations to include basic MI, counterterrorism, counterdrugs, medical intelligence, WMD and proliferation, United Nations (UN) peacekeeping and multinational support, missile and space intelligence, noncombatant evacuation operations (NEOs), targeting, combat assessment, and battle damage assessment (BDA).

National Security Agency (NSA)

2-29. NSA ensures cryptologic planning and support for joint operations. Working with the tactical cryptologic units of a command, the NSA provides SIGINT and information security (INFOSEC), encompassing communications security (COMSEC) and computer security, as well as telecommunications support and operations security (OPSEC). The people and equipment providing SIGINT, INFOSEC, and OPSEC constitute the United States Cryptologic System (USCS). The NSA, through the USCS, fulfills cryptologic command and/or management, readiness, and operational responsibilities in support of military operations according to the Secretary of Defense tasking, priorities, and standards of timeliness.

National Geospatial-Intelligence Agency (NGA)

2-30. The NGA (formerly National Imagery and Mapping Agency [(NIMA)] mission is to provide timely, relevant, and accurate intelligence and geospatial information in support of national security objectives of the United States. The Director of NGA advises the Secretary of Defense, Director Central Intelligence (DCI), Chairman of the Joint Chiefs of Staff (JCS), and the combatant commanders on imagery, IMINT, and geospatial information. The Operations Directorate, Customer Support Office, is the focal point for interface with external customers, including the JCS, combatant commands, services, and national and defense agencies.

National Reconnaissance Office (NRO)

2-31. The NRO develops and operates unique and innovative space reconnaissance systems and conducts intelligence-related activities essential for US security. The role of the NRO is to enhance US government and military information superiority across full spectrum operations. NRO responsibilities include supporting I&W, monitoring arms control agreements, and performing crisis support to the planning and conduct of military operations. The NRO accomplishes its mission by building and operating IMINT and SIGINT reconnaissance satellites and associated communications systems

US Navy (USN)

2-32. Naval intelligence products and services support the operating naval forces, the Department of the Navy, and the maritime intelligence requirements of national level agencies. Naval intelligence operates the National Maritime Intelligence Center (NMIC). Naval intelligence responsibilities include maritime intelligence on global merchant affairs, counter-narcotics, fishing issues, ocean dumping of radioactive waste, technology transfer, counter-proliferation, cryptologic related functions, CI, I&W support, management of Coast Guard collection, and development of new weapons systems and countermeasures.

US Marine Corps (USMC)

2-33. USMC intelligence provides pre-deployment training and force contingency planning for requirements that are not satisfied by theater, other service, or national capabilities. The Marine Corps Intelligence Agency (MCIA) handles the integration, development, and application of GMI, technical information, all-source production, and open-source information.

US Air Force (USAF)

2-34. USAF ISR assets fill a variety of roles to meet US national security requirements. The USAF operates worldwide ground sites and an array of airborne ISR platforms to meet national level intelligence requirements. To support day-to-day USAF operations and to meet specific USAF requirements, intelligence professionals at the wing and squadron levels use suites of interoperable analysis tools and dissemination systems to tailor information received from all levels and agencies in the intelligence community. USAF responsibilities include all-source information on aerospace systems and potential adversaries' capabilities and intentions, cryptologic operations, I&W, IO, and criminal investigative and CI services.

NON-DOD AGENCIES

2-35. Although the primary focus of non-DOD members of the intelligence community is strategic intelligence and support to the President and the Secretary of Defense, these agencies also produce intelligence and intelligence products that support operational and tactical ARFOR. This responsibility includes assessing potential issues and situations that could impact US national security interests and objectives. These agencies identify global and regional issues and threats. Some of the intelligence products and services these agencies provide are essential to accurate assessment of the threat and battlespace, particularly during stability operations and support operations.

Central Intelligence Agency (CIA)

2-36. The CIA's primary areas of expertise are in HUMINT collection, imagery, all-source analysis, and the production of political and economic intelligence. CIA and military personnel staff the CIA's Office of Military Affairs (OMA). As the CIA's single POC for military support, OMA negotiates, coordinates, manages, and monitors all aspects of agency support for military operations. This support is a continuous process, which the agency enhances or modifies to respond to a crisis or developing operation. Interaction between OMA and the DCI representatives to the Office of the Secretary of Defense (OSD), the Joint Staff, and the combatant commands facilitates providing national level intelligence in support of joint operations, contingency and operations planning, and exercises.

Department of State, Bureau of Intelligence and Research

2-37. The Bureau of Intelligence and Research coordinates programs for intelligence, analysis, and research; it produces intelligence studies and current intelligence analyses essential to foreign policy determination and execution. Its subordinate Bureau of International Narcotics Matters develops, coordinates, and implements international narcotics control assistance activities. It is the principal POC and provides policy advice on international narcotics control matters for the Office of Management and Budget (OMB), the NSC, and the White House Office of National Drug Control Policy (ONDCP). The Bureau also oversees and coordinates the international narcotics control policies, programs, and activities of US agencies.

Department of Energy, Office of Nonproliferation and National Security

2-38. The Department of Energy's Office of Nonproliferation and National Security assists in the development of the State Department's policy, plans, and procedures relating to arms control, nonproliferation, export controls, and safeguard activities. Additionally, this office is responsible for—

- Managing the department's R&D program.
- Verifying and monitoring arms implementation and compliance activities.
- Providing threat assessments and support to headquarters and field offices.

Department of the Treasury

2-39. The US Treasury Department's intelligence-related missions include producing and disseminating foreign intelligence relating to US economic policy and participating with the Department of State in the overt collection of general foreign economic information.

Federal Bureau of Investigation (FBI)

2-40. The FBI is the principal investigative arm of the Department of Justice (DOJ) and has primary responsibility for CI and counterterrorism operations conducted in the US. CI operations contemplated by any other organizations in the US must be coordinated with the FBI. Any overseas CI operation conducted by the FBI must be coordinated with the CIA.

Department of Homeland Security

2-41. The mission of the Department of Homeland Security is to develop and coordinate the implementation of a comprehensive national strategy to secure the US from terrorist threats or attacks. The organizational construct of homeland security mission is a framework of prepare, deter, preempt, defend, and respond. Component agencies will analyze threats and intelligence, guard our borders and airports, protect our critical infrastructure, and coordinate the response of our nation for future emergencies. The component agencies of the department include—

- US Coast Guard (USCG).
- US Customs Service.
- US Border Patrol.
- Federal Emergency Management Agency (FEMA).
- Immigration and Naturalization Service (INS).
- Transportation Security Administration.
- Federal Protective Service.
- Office of Domestic Preparedness.

United States Coast Guard (USCG)

2-42. The USCG, subordinate to the Department of Homeland Security, has unique missions and responsibilities as both an armed force and a law enforcement agency (LEA), which makes it a significant player in several national security issues. The USCG intelligence program supports counter-drug operations, mass seaborne migration operations, alien migration interdiction operations, living marine resource enforcement, maritime intercept operations, port status and/or safety, counterterrorism, coastal and harbor defense operations, and marine safety and/or environmental protection.

Other Agencies

2-43. There are a number of US Government agencies and organizations, not members of the intelligence community, that are responsible for collecting and maintaining information and statistics related to foreign governments and international affairs. Organizations such as the Library of Congress, the Departments of Agriculture and Commerce, the National Technical Information Center, US Information Agency, US Information Service, and the US Patent Office are potential sources of specialized information on political, economic, and military-related topics. The intelligence community may draw on these organizations to support and enhance research and analysis and to provide relevant information and intelligence for planners and decisionmakers. Many other US Government agencies may become directly involved in supporting DOD especially during stability operations and support operations. (See JP 2-02 for a description of agency support to joint operations and intelligence.) These organizations include—

- Department of Transportation.
- Disaster Assistance Response Team within the Office of Foreign Disaster.
- US Agency for International Development.
- NGA.

UNIFIED ACTION INTELLIGENCE OPERATIONS

2-44. In unified action, ARFOR synchronize their actions with those of other participants to achieve unity of effort and to accomplish the combatant commander's objectives. Unified action links subordinates to the combatant commander under combatant command (command authority) (COCOM). Multinational, interagency, and nonmilitary forces work with the combatant commander through cooperation and coordination. Combatant commanders form theater strategies and campaigns, organize joint forces, designate operational areas, and provide strategic guidance and operational focus to subordinates. The aim is to achieve unity of effort among many diverse agencies in a complex environment. Subordinate JFCs synchronize joint operations in time and space, direct the action of foreign military forces (multinational operations), and coordinate with governmental and NGOs (interagency coordination) to achieve the same goal.

2-45. The J2 staff provides intelligence promptly, in an appropriate form, and by any suitable means to those who need it. Intelligence personnel ensure that the consumers understand the intelligence and assist them as they apply the intelligence to their operations.

2-46. Dissemination requires establishing appropriate communications systems and procedures. The J2 and other intelligence personnel must fully participate in all operation planning and execution, and develop close working relationships with the JFC and other staff elements.

2-47. The commander and staff assess intelligence operations to determine their effectiveness and to make any necessary improvements. The intelligence process functions presented in Chapter 4 provide the criteria for evaluating intelligence operations.

JOINT INTELLIGENCE OPERATIONS

2-48. The JTF commanders and his intelligence staff must—

- Understand the intelligence requirements of superior, subordinate, and component commands.
- Identify organic intelligence capabilities and shortfalls.
- Access theater and national systems to ensure appropriate intelligence and CI products are available to the JTF.

2-49. The JTF's intelligence effort focuses on integrating multi-source information and multi-echelon intelligence into all-source intelligence products that provide clear, relevant, and timely knowledge of the enemy and battlespace. These products must be in formats that are readily understood and directly usable by the recipient in a timely manner. They must neither overload the user nor the communications architecture.

2-50. The J2 is the JTF commander's focal point for intelligence. The J2 directly supports the JFC's responsibilities for determining objectives, directing operations, and evaluating the effects of those operations. The J2 supports the execution of the plan with the intelligence needed to sustain the operation, attain joint force objectives, provide support to subordinate commands, and continually support FP efforts. The J2 analyzes the potential threat situation and provides assessments to support friendly opportunities. The J2 then supports the execution of the plan with the operational intelligence needed to sustain the operations, attain joint force objectives, and support FP. To maintain the initiative, the JFC will seek to get inside the adversary's decisionmaking cycle; that is, the JFC will seek to develop procedures and an organization in order to receive new and accurate intelligence and respond to the new situation faster than the adversary. The J2 must help in identifying the adversary's decisionmaking cycle and identifying vulnerabilities that may be exploited. The J2 also—

- Ensures the provision of the required ISR support to the JTF and its subordinate functional and service components.
- Assists the JTF commander in defining intelligence responsibilities and PIRs.
- Actively participates in joint staff planning and in planning, preparing, executing, and assessing ISR efforts. This includes leading the JTF's joint collection board and providing representation at the joint targeting board.

2-51. Figure 2-3 shows a typical JTF J2 organization. The overall organization of the JTF and operations will dictate actual composition of the J2. At a minimum, a core element of analytical, ISR management, and administrative capabilities is required.

FM 2-0

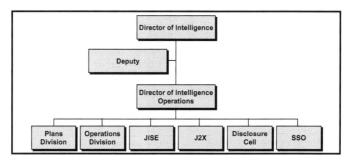

Figure 2-3. Typical Joint Task Force J2 Organization.

Considerations In Joint Intelligence Operations

2-52. The Army, Air Force, Navy, Marine Corps, and SOF must provide globally responsive assets to support the combatant commanders' theater strategies and the national security strategy. The capabilities of the other armed forces complement those of ARFOR. During joint operations, they provide support consistent with JFC-directed missions. When conducting joint intelligence operations, there are a number of unique problems that can arise due to the complexity of integrating the efforts of the different services and commands. Elements affecting joint intelligence operations among the different services include the following:

- Intelligence liaison is critical to the success of intelligence operations and requires early establishment, particularly between units that have not routinely trained together and possess differing capabilities. As a minimum, organizations exchange liaison teams with the higher echelon organization. Additional liaison may be necessary to facilitate joint force collection, production, or dissemination requirements. Liaison teams—
  - Support planning and C2 of intelligence operations.
  - Ensure timely two-way flow of intelligence between commands.
  - Manage intelligence and resource requirements of the subordinate command.
  - Advise the commander on service ISR capabilities, limitations, and employment.
- Commanders and staffs use IPB to understand the battlespace and develop or refine plans and orders. IPB products exchanged between echelons ensure a common picture of the battlespace and estimate of the situation.
- Communications considerations for joint operations include—
  - Planning for intelligence communications transition to facilitate execution of branches or sequels to the plan or to accommodate shifting of the main effort from one force to another.

- Identifying the initial communications architecture to include establishing procedures and protocols for information exchanges (databases, text, imagery, voice, and video).
- Balancing the availability of service-unique intelligence systems between echelons or services. This may require each service providing additional resources. The senior commander is responsible for allocating resources.
- Disseminating intelligence between commands and services. Additional communications equipment, intelligence terminals, and personnel may be required to balance capabilities between services.
- Identifying the databases each service possesses or has access to; determining which databases will support the operation and, if necessary, merge them into a single database; and ensure access by the entire force.
- Providing a focal point for subordinate command access to national or joint intelligence is essential. The senior commander will request and allocate resources required to support this access.

2-53. The JFC's intelligence requirements, concept of operation, and intent drive the ISR effort. The different organizations and services participating in joint intelligence operations must continuously share information, intelligence, and products to satisfy requirements. See FM 2-01 for details on intelligence requirements and RM. Activities required to facilitate an effective joint collection effort include—

- Coordinating intelligence and ISR operations to optimize capabilities of collection assets and reduce duplication of effort.
- Integrating supporting national and theater intelligence collection assets into the intelligence synchronization plan.
- Establishing procedures for tracking and handing off HPTs between services and echelons.
- Establishing procedures for cueing Army and other service collection assets.
- Coordinating airspace for ISR assets.
- Maximizing available linguist capabilities. Shortages of military linguists trained in target languages may require cross-leveling Army and other service linguists.

2-54. Reporting and intelligence production considerations in joint operations include—

- Establishing production criteria and thresholds that produce timely and relevant intelligence keyed to the commander's intelligence and targeting requirements. The ISM, attack guidance matrix (AGM), and HPT list are examples of tools used to support joint intelligence and targeting efforts. The sharing of these products to all echelons and services is crucial.
- Establishing common methodology and criteria for producing the BDA and supporting the combat assessment function.
- Establishing set standards for the number and frequency of periodic reports such as intelligence summaries (INTSUMs).

Joint Intelligence Architecture

2-55. In addition to the J2 staffs at every joint level of command, the key organizations in the joint intelligence architecture are the National Military Joint Intelligence Center (NMJIC) (see Figure 2-4), the Joint Intelligence Centers (JICs) or Joint Analysis Centers (JACs) of the unified commands, and, when formed, the JTF's joint intelligence support element (JISE). Working together, these organizations play the primary role in managing and controlling joint intelligence operations. The formal relationships which link these organizations facilitate information management and optimize complementary intelligence functions by echelon without obstructing the timely flow of intelligence up, down, or laterally.

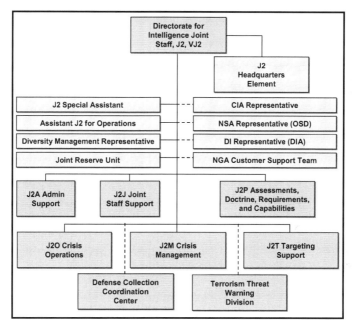

Figure 2-4. National Military Joint Intelligence Center.

2-56. The NMJIC is the focal point for intelligence activities in support of joint operations. The NMJIC is the CJCS J2 intelligence watch within the National Military Command Center (NMCC) in the Pentagon. When the assigned or attached assets cannot satisfy the combatant commander and the JTF commander's crisis-related and time-sensitive intelligence requirements, the NMJIC, as part of the J2's staff, sends tasks to appropriate agencies to fill the requirements. During crises, the NMJIC expands as necessary to establish a working group, intelligence task force, or an expanded intelligence task force. The NMJIC participates in targeting by developing national level target lists. When requested, the NMJIC supports the theater in performing BDA. Besides supporting the combatant commands and JTFs, the NMJIC supports any multinational partners and prescribed international organizations.

2-57. The JIC is organized in accordance with the combatant commander's prerogatives, but normally performs the general functions described in JP 2-0 and specific unified command intelligence TTP. The JIC is responsible for providing intelligence to JTF and its subordinates during military operations. If the JIC cannot meet the combatant commander's requirements, the JIC forwards requests to the NMJIC or to subordinate command levels through established channels, using standard command procedures. In some cases, the JIC may also ensure timely support by approving a direct communication path, in advance, between requesters such as the JTF JISE and outside producers. In this case, the parties must inform the JIC of all requests as they occur. This method is most appropriate when the parties require products and services, which the JIC does not routinely produce.

2-58. At the discretion of the JTF commander, the J2 can establish a JISE during the crisis or preparation stage for operations to augment the J2 staff. The JISE is a tailored subset of the JIC or the intelligence organization of the service component designated as the JTF headquarters.

2-59. Under the direction of the J2, the JISE normally manages the intelligence collection, production, and dissemination for the JTF. The JISE provides intelligence to JTF operational forces and performs common intelligence functions. By design, the JISE is scaleable and can expand to meet the needs of the JTF and the operating environment. It is composed of analytical experts, analysis teams, and ISR managers that provide services and products, which the JTF, JTF staff, and subordinate components require. These experts, mentioned above, focus on solving the operational intelligence problems of concern to the JTF commander. The JISE's capability to perform all-source analysis and ISR synchronization is key to producing operational intelligence that is timely, relevant, and complete. Figure 2-5 illustrates the features of a typical JISE.

Figure 2-5. Typical Joint Intelligence Support Element.

Joint Task Force Intelligence Organizations

2-60. In addition to the JISE, the JTF commander and J2 may require other supporting JICs or teams based on projected operations. The JTF commander may make a request to the NMJIC for specific national intelligence agency capabilities. The NMJIC evaluates and coordinates the JTF commander's requirements with the J3, J5, and national intelligence agencies and tailors the composition of the deployment packages to meet those needs. The deployment packages, such as the National Intelligence Support Team (NIST), provide access to the entire range of capabilities resident in the national intelligence agencies and can focus those capabilities on the JTF commander's intelligence requirements. The J2X manages and coordinates the HUMINT and CI activities of national, theater, and service components operating within the JTF's JOA. The Joint Captured Materiel Exploitation Center (JCMEC) assists in management of recovery, exploitation, and disposal of captured enemy equipment (CEE). The JTF commander's requirements dictate the composition and tailoring of such deployment packages.

Augmentation Considerations

2-61. Depending on the scale of the operations, the intelligence organizations described above and those of the JTF's subordinate command may require personnel augmentation. Optimum use of available intelligence assets is essential to ensure quality support in meeting the JTF commander's requirements.

2-62. The demand for additional intelligence increases significantly during crisis and wartime operations. As the need arises for more intelligence personnel, the intelligence presence also increases at all command levels. Locating additional intelligence personnel and knowing how to integrate those personnel into the operation is vital. The JTF J2 should identify intelligence personnel augmentation requirements in accordance with the Chairman, Joint Chiefs of Staff (CJCS) Instruction 1301.01. The combatant commander's joint table of mobilization and distribution (JTMD) also contains the standing augmentation requirements. The JTMD should reflect the need for either IMA or Individual Ready Reserve (IRR) personnel. The combatant commander and the JTF refine personnel requirements and initiate requests when they anticipate or start an operation.

2-63. A consideration for the JTF when requesting support or augmentation is that these national level teams and individual augmentees are not totally self-contained elements; rather they require logistic, information, and other support from the supported command. Each deployment is unique based on mission, duration, team composition, and capabilities required. A full NIST, for example, requires a private, access-controlled area within a sensitive compartmented information facility (SCIF) work environment and dedicated secure communications.

2-64. For more information on intelligence operations as they apply to other armed services, see the individual service intelligence doctrine. See also JP 2-0 series, JP 3-0, JP 3-55, and JP 5-0 for more details on joint intelligence operations and considerations.

Multinational Intelligence

2-65. Multinational intelligence operations take place within the structure of an alliance or coalition. Some multinational military organizations, such as the North Atlantic Treaty Organization (NATO) and the UN Command in the Republic of Korea (ROK), are highly structured and enduring. Others, such as the coalition formed during the Gulf War, are less formal and temporary.

2-66. In multinational operations, the multinational force commander (MNFC) exercises command authority over a military force composed of elements from two or more nations. Therefore, in most multinational operations, the JTF must share intelligence, as necessary, for mission accomplishment with foreign military forces and coordinate exchange of intelligence with those forces.

2-67. In some circumstances, the JTF may need to seek authority to go outside the usual political-military channels to provide information to NGOs. The JTF must tailor intelligence policy and dissemination criteria to each multinational operation.

INTELLIGENCE REACH

2-68. Military forces use intelligence reach to rapidly access information, receive support, and conduct collaboration and information sharing with other units and organizations unconstrained by geographic proximity, echelon, or command.

2-69. Intelligence reach supports minimizing the deployed footprint of ISR assets. By providing enhanced information and tailored intelligence products, intelligence reach can greatly enhance the intelligence capabilities of the unit and play a significant role in improving the commander's ability to make decisions on the battlefield. It allows the commander to harness national, joint, foreign, and other military organization resources. Table 2-1 shows examples of partners and sources for intelligence reach.

2-70. Detailed planning and training are critical to the success of intelligence reach operations. The following are steps that the staff can take to ensure optimal use, operability, and effectiveness of intelligence reach:

- Establish data exchange methods and procedures.
- Establish electronic message transfer procedures.
- Establish homepages for identified forces.
- Establish POCs for I&W centers, production centers, Theater JICs, DIA, INSCOM, and their major subordinate commands such as NGIC and the theater intelligence brigades and groups.
- Ensure the intelligence staff has the necessary personnel, training, automated systems, bandwidth, and resources to conduct intelligence reach.
- Determine IRs through staff planning. Develop production requirements for identified intelligence gaps.
- Order geospatial products for the projected joint AOI.
- Establish and maintain a comprehensive directory of intelligence reach resources before deployment and throughout operations. The value of intelligence reach will greatly increase as the staff develops and maintains ready access to rich information resources. These resources are numerous and may include, for example, Army, Joint, DOD, non-DOD, national, commercial, foreign, and university research programs. Know what types of information the resources can provide. Continuously expand the resource directory through identification of new resources.

- Use intelligence reach first to fill intelligence gaps and requirements and answer RFIs. This technique can preclude unnecessary tasking or risk to limited ISR assets.
- Maintain continuous situational understanding and anticipate intelligence requirements. Use intelligence reach to fulfill these requirements and provide the results to the commander and staff for planning and decisionmaking.
- Exchange intelligence reach strategies with other units.
- Present the information retrieved through intelligence reach in a usable form. Share the information derived from intelligence reach with subordinate, lateral, and higher echelons. Ensure follow-on forces have all information as well.

Table 2-1. Examples of Intelligence and Reach Partners and Sources.

ARMY	SERVICES	JOINT	DOD
ACE ISE TSE (TIB/TIG) 902d NGIC ATCAE ARISCs	ONI NMIC AIA NAIC MCIA	USEUCOM JAC USSOUTHCOM JIC USSOCOM JIC USSPACECOM CIC USJFCOM AIC USCENTCOM JIC USTRANSCOM JIC USSTRATCOM USPACOM JIC USNORTHCOM JIC	DIA CMO MSIC AFMIC DHS NGA NSA RSOCs
NON-DOD	NATIONAL	COMMERICAL	FOREIGN
DOE FBI DOS DEA FEMA INS	DCI NIC CIA NRO INR DOT, Office of Intel Support	RAND Jane's Defence Weekly Economic Intelligence Unit CNN Reuters Associated Press United Press International	DIS NFHQ DIO

INTELLIGENCE REACH COMPONENTS

2-71. Intelligence reach requires the G2/S2 to develop a strategy on how best to support the unit's mission with intelligence reach capabilities. There are six basic components of the strategy:

- Push
- Database Access.
- Pull.
- Broadcast Services.
- Collaborative Tools.
- Requirements Management.

Push

2-72. Push occurs when the producers of intelligence or information are knowledgeable of the customer's requirements and are able to send the desired intelligence to the customer without further requests. Push is accomplished through the most appropriate or efficient dissemination means. Unless coordinated as an acceptable dissemination method, push is not accomplished solely by posting intelligence products on a web page. The entity that posts the document must ensure the intended recipient is aware of the product's location or has received the product.

2-73. Push begins with the statement of intelligence interest (SII). The SII establishes the unit's intelligence and IRs. The SII is prepared by intelligence staff organizations to register their interest in receiving recurring hardcopy and softcopy reports, studies, and publications covering a wide variety of intelligence subjects. The J2/G2/S2 works with the Department of the Army Production and Dissemination Management (DAPDM) to establish a profile for information transfer. Intelligence reports or information meeting the unit's requirements are then automatically sent directly to the unit's classified or unclassified networks or communication systems.

2-74. Organizations that push information and data down to subordinate units must be careful of the amount, detail, and focus of the information they are sending. Too much information will overwhelm the subordinate unit. Crucial information may be lost in the midst of an overabundant flow of information, much of which is of little use. Intelligence reach often runs more efficiently if the user pulls the information or at least focuses the intelligence producers to send the appropriate information with the correct level of detail at the appropriate time.

Database Access

2-75. Access to local, theater, DOD, non-DOD, and commercial databases allows analysts to leverage stored knowledge on topics ranging from basic demographics to OB information. A validated DIA Customer Number (acquired by the J2/G2/S2) in combination with Secret Internet Protocol Network (SIPRNET) and Joint Worldwide Intelligence Communications System (JWICS) connectivity establishes access to most of the databases online. The challenge for an analyst is to gain an understanding of the structure, contents, strengths, and weaknesses of the database regardless of database type. Additionally, the procedures are often difficult for extracting portions or downloading and transferring the database to unit level automated information systems.

2-76. Each intelligence discipline has unique databases established and maintained by a variety of agencies. Database access is accomplished through unit or agency homepages via SIPRNET (Intelink-S) and JWICS (Intelink). The DAPDM office, upon approving the unit's SII, validates the requirement for access to the majority of these databases. Units coordinate with the DA dissemination and program manager (DPM) and/or the agency for access to those databases requiring passwords and permissions beyond normal Intelink-S and Intelink access.

Pull

2-77. Pull occurs when the requestor is familiar enough with existing databases and products to be able to anticipate the location of the desired information. Knowledge of both the types and locations of intelligence databases can greatly increase the efficiency of the Intelligence BOS by saving time and effort on the part of analysts at every echelon.

2-78. The G2/S2 must also ensure that intelligence reach capability extends to multinational forces, augmentees, and other services or organizations working as part of the JTF. The G2/S2 must forward relevant intelligence to units which do not possess the necessary automation to conduct intelligence reach.

Broadcast Services

2-79. Broadcast services are an integrated, interactive dissemination system, focusing on tactical user's IRs using joint message formats. The theater broadcast's data prioritization and processing occur through information management. Selected tactical, theater, DOD, and national collectors broadcast their messages using prescribed message formats. The analyst decides which types of messages are required. Broadcast service data is responsive to the operational commander's information needs and allows tactical users to construct successively detailed pictures of the battlespace.

2-80. Tactical users require intelligence and information to be available at the lowest classification level possible (for example, collateral SECRET) and releasable, as necessary, to multinational partners. Broadcast service information management will adhere to national policy as it relates to the provision of data to NATO or multinational partners.

Collaborative Tools

2-81. Collaboration is the sharing of knowledge, expertise, and information, normally online, and may take numerous forms. Collaborative tools are computer-based tools (groupware) that help individuals work together and share information. They allow for virtual online meetings and data sharing. As an example of the use of collaborative tools, the President or the Secretary of Defense and DOD during a crisis situation will establish a number of crisis action teams (CATs) or joint interagency working groups (JIAWGs). These groups or teams are formed to address specified subjects or topics in support of the warfighter or decisionmaker. The groups will focus on the crisis and normally publish their products on a homepage. Analysts with online access can participate in a number of ways. This includes passively accessing the homepages to study the products, sending queries to the identified POCs, or by having their organization join the CAT or JIAWG. Table 2-2 shows some examples of collaborative tools.

Requirements Management

2-82. The RM system provides a mechanism for submitting RFIs, tasking, and managing ISR assets. Analysts who are trained and familiar with the RM process and the various tasking procedures can leverage its systems for refined information. Each intelligence discipline has established procedures for requesting specific information. It is therefore advantageous to have someone familiar with each discipline's procedures to participate in synchronizing the ISR effort. For example, an analyst receives a HUMINT-based intelligence information report, which provides information on an event or subject. If he requires more refined data or clarification, he should submit a time-sensitive requirement or a source-directed requirement. This creates a two-way communication from the field to the collectors without creating additional standing requirements. The more familiar the analyst is with the RM process, the better he can leverage it.

2-83. The intelligence reach component of RM includes the ability of an intelligence officer at any level to request information, which is beyond what is available at his location. The normal procedure for obtaining intelligence or information not obtainable through the use of available ISR assets is to submit an RFI to the next higher echelon. Users enter RFIs into the RM system. See ST 2-33.5 for more information on intelligence reach operations and FM 2-01 for more information on RM.

Table 2-2. Examples of Collaborative Tools.

Tool	Description
Chat (Audio and Text)	Used to conduct conversations online.
Whiteboard	Permits real-time display of drawings, pictures, or documents for group discussion and comment. Participants can annotate in real time as well.
Bulletin Board	Used to post notices and facilitate discussions on any topic.
Video	Used at a desktop computer or a video teleconferencing (VTC) center to allow the person or group to see with whom they are communicating.
Discussion Groups (Newsgroups)	Topics are posted to a website for discussion and comment where participants can follow a line of discussion on a topic.
File-Sharing Tools	Virtual file cabinets allow information to be stored on web servers, and are available to anyone having access to the site and electronic permission to use the files.
Presentation Tools	Used in a virtual auditorium to allow lectures and briefings to be given to an audience.
Application Sharing	An entire team can use an application running on one computer to revise documents.
Text Tools	Allows live text input and editing by group members. Once complete, the text document can be copied into word processing software.
E-mail	Electronic mail.
Persistent Capability	The ability to preserve files, briefings, or other team or project material for future reference. Properly organized, it becomes an information management device and is invaluable to a long-term effort.
Instant Messaging	This allows real-time exchange of notes and messages.

PART TWO

Intelligence in Full Spectrum Operations

Part Two provides a primer of what and who constitutes both MI and the Intelligence BOS and the process that these entities use in order to provide warfighters and decisionmakers with the intelligence they require in order to accomplish their respective missions.

Chapter 3 discusses the role of MI and the Intelligence BOS within full spectrum operations. It provides an overview of intelligence readiness, particularly the intelligence requirements associated with force projection. The concept of MI asset technical control is discussed as a complement to, not a replacement of, the Army's command and support relationships.

Chapter 4 presents the Intelligence Process—the Intelligence BOS methodology that accomplishes the primary focus of intelligence in full spectrum operations, which provides the warfighter with effective intelligence.

Chapter 3

Fundamentals in Full Spectrum Operations

FULL SPECTRUM OPERATIONS

3-1. Army commanders at all echelons may combine different types of operations simultaneously and sequentially to accomplish missions. Full spectrum operations include offensive, defensive, stability, and support operations. Missions in any environment require ARFOR that are prepared to conduct any combination of these operations:

- **Offensive operations** aim to destroy or defeat an enemy. Their purpose is to impose US will on the enemy and achieve decisive victory.
- **Defensive operations** defeat an enemy attack, buy time, economize forces, or develop conditions favorable for offensive operations. Defensive operations alone normally cannot achieve a decision. Their purpose is to create conditions for a counteroffensive that allow ARFOR to regain the initiative.
- **Stability operations** promote and protect US national interests by influencing the threat, political, and information dimensions of the OE through a combination of peacetime developmental, cooperative activities, and coercive actions in response to crises. Regional security is supported by a balanced approach that enhances regional stability

and economic prosperity simultaneously. ARFOR presence promotes a stable environment.
- **Support operations** employ ARFOR to assist civil authorities, foreign or domestic, as they prepare for or respond to crises and relieve suffering. Domestically, ARFOR respond only when the President or Secretary of Defense direct. ARFOR operate under the lead federal agency and comply with provisions of US law, to include the Posse Comitatus and Stafford Acts. (See FM 3-0.)

3-2. Intelligence supports the commander across full spectrum operations. It helps the commander decide when and where to concentrate sufficient combat power to overwhelm the enemy. ISR is essential for the commander to preclude surprise from the enemy, maintain the initiative on the battlefield, and win battles. Commanders at all levels synchronize intelligence with the other BOSs to maximize their ability to see and strike the enemy simultaneously throughout the AO.

3-3. Every soldier in the command is responsible for detecting and reporting enemy activities, dispositions, and capabilities. This task is critical because the environment we operate in is characterized by violence, uncertainty, complexity, and asymmetric methods by the threat. The increased situational awareness that soldiers develop through personal contact and observation is a critical element of that unit's ability to more fully understand the OE. However, soldiers collect information, they are not intelligence collectors. While medical personnel cannot be assigned ISR tasks due to their Geneva Convention category status, medical personnel who gain information through casual observation of activities in plain view while discharging their humanitarian duties will report the information to their supporting intelligence element.

OFFENSIVE OPERATIONS

3-4. Offensive operations at all levels require effective intelligence to help the commander avoid the enemy's main strength and to deceive and surprise the enemy. During offensive operations, intelligence must provide the commander with the composition, disposition, limitations, employment characteristics, and anticipated enemy actions in a timely enough manner for the commander to significantly affect the enemy commander's decision cycle. It ensures commanders have the intelligence they need to conduct offensive operations with minimum risk of surprise.

DEFENSIVE OPERATIONS

3-5. The immediate purpose of defensive operations is to defeat an enemy attack. Commanders defend to buy time, hold key terrain, hold the enemy in one place while attacking in another, or destroy enemy combat power while reinforcing friendly forces. Intelligence should determine the enemy's strength, COAs, and location of enemy follow-on forces. The defending commander can then decide where to arrange his forces in an economy-of-force role to defend and shape the battlefield. Intelligence support affords him the time necessary to commit the striking force precisely.

3-6. Intelligence in area defensive operations identifies, locates, and tracks the enemy's main attack and provides the commander time to allocate sufficient combat power to strengthen the defense at the point of the enemy's main effort. Intelligence should also identify where and when the commander can most decisively counterattack the enemy's main effort or exploit enemy vulnerabilities.

3-7. Although the battlefield is normally organized as decisive, shaping, and sustaining operations, commanders conducting defensive operations may delineate intelligence in battlefield organizational terms of deep area, close area, and rear area.

STABILITY OPERATIONS

3-8. The environment is often much more complex during stability operations and as a result intelligence is often more complex. In fact, intelligence is even more important a factor (or operational multiplier) during stability operations. As a result, the commander must be even more involved in and knowledgeable of ISR (to include ISR operations the commander controls and other higher level ISR operations that may be occurring within his AO) during stability operations. For example, the commanders must understand the complex details of HUMINT and special access program (SAP) operations.

3-9. The commander requires the appropriate intelligence and IPB products in order to determine how best to influence the threat, political and information dimensions of the operational environment, and enhance regional stability. The identification and analysis of characteristics of the terrain and weather, politics, infrastructure, health status, civilian press, attitudes, and culture of the local populace and all possible threats are important in conducting stability operations. A lack of knowledge concerning local politics, customs, and culture could lead to US actions which attack inappropriate targets or which may offend or cause mistrust among the local population.

SUPPORT OPERATIONS

3-10. Support operations are usually nonlinear and noncontiguous. Commanders designate the decisive, shaping, and sustaining operations necessary for mission success. However, determining the intelligence requirements that drive the ISR effort intending to identify any potential threat's centers of gravity and decisive points may require a more complex and unorthodox thought process than that used in offensive and defensive operations. The G2/S2 may have to define the enemy differently. In support operations, the adversary is often the effects of disease, hunger, or disaster on a civilian population. US forces conducting support operations must also fully understand the organization and identity of key figures or groups within the region where they are operating, as these figures may influence greatly the actions of the population—both civilian and military.

ELEMENTS OF COMBAT POWER

3-11. Combat power is the ability to fight. It is the total means of destructive or disruptive force, or both, that a military unit or formation can apply against the adversary at a given time. Commanders combine the elements of combat power to meet constantly changing requirements and defeat an enemy. The elements of combat power are—

- **Manuever** is the employment of forces, through movement combined with fire or fire potential, to achieve a position of advantage with respect to the enemy to accomplish the mission. Maneuver is the means by which commanders concentrate combat power to achieve surprise, shock, momentum, and dominance.
- **Firepower** provides the destructive force essential to overcoming the enemy's ability and will to fight. Firepower magnifies the effects of maneuver by destroying enemy forces and restricting his ability to counter friendly actions; maneuver creates the conditions for the effective use of firepower.
- **Leadership** is the most dynamic element of combat power. Confident, audacious, and competent leadership focuses on the other elements of combat power and serves as the catalyst that creates conditions for success. Leadership provides purpose, direction, and motivation in all operations.
- **Protection** is the preservation of fighting potential of a force so the commander can apply maximum force at the decisive time and place. Protection is neither timidity nor risk avoidance.
- **Information** enhances leadership and magnifies the effects of maneuver, firepower, and protection. Today, ARFOR use information collected to increase their situational understanding before engaging the enemy. Information from the COP and running estimates, transformed into situational understanding, allows commanders to combine the elements of combat power in new ways.

THE FOUNDATIONS OF ARMY OPERATIONS

3-12. Understanding the principles of war and tenets of Army operations is fundamental to operating successfully across the range of military operations. The principles of war and tenets of Army operations form the foundation of Army operational doctrine. (Refer to FM 3-0 for a full description of the Principles of War and Tenets of Army Operations.)

- The **Principles of War** provide general guidance for conducting war and military operations other than war at the strategic, operational, and tactical levels. The principles are the enduring bedrock of Army doctrine.
- The **Tenets of Army Operations** build upon the principles of war. They further describe the characteristics of successful operations. These tenets are essential to victory.

THE OPERATIONAL FRAMEWORK

3-13. The operational framework consists of the arrangement of friendly forces and resources in time, space, and purpose with respect to each other and the enemy or situation. It consists of the **AO, battlespace,** and **battlefield organization.** The framework establishes an area of geographic and operational responsibility and provides a way for commanders to visualize how to employ forces against the enemy. Commanders design an operational framework to accomplish their mission by defining and arranging its three components. They use the framework to focus combat power. The operational framework provides a mechanism through which the commander can focus the Intelligence BOS effort. Understanding the operational framework, AOIR, and intelligence coordination line (ICL) and their relationship to each other is key to planning and executing ISR operations.

- AO is an operational area defined by the JFC for land and naval forces. AOs do not typically encompass the entire operational area of the JFC but should be large enough for component commanders to accomplish their missions and protect their forces. AOs should also allow component commanders to employ their available systems to the limits of their capabilities. The AO is the basic control measure for all operations that defines the geographical area for which a particular unit is responsible. The commander—
  - Assumes responsibility for intelligence, maneuver, fires, terrain management, security, and movement within his AO.
  - Establishes control measures within his AO to assign responsibilities, coordinate intelligence, fires, and maneuver, and to control other activities.
- Battlespace is the environment, factors, and conditions that must be understood to successfully apply combat power, protect the force, or complete the mission. This includes air, land, sea, space, and the included enemy and friendly forces; facilities; weather; terrain; the EMS; and the information environment within the operational areas and AOIs. The G2/S2 performs IPB and synchronizes ISR activities throughout the battlespace as determined by the commander's METT-TC considerations.
  - Area of influence is the geographical area in which a commander can directly influence operations by maneuver of FS systems normally under the commander's command or control. Areas of influence surround and include the associated AO.
  - AOI is that area of concern to the commander, including the area of influence, areas adjacent thereto, and extending into enemy territory to the objectives of current or planned operations. This area also includes areas occupied by enemy forces who could jeopardize the accomplishment of this mission. (See JP 1-02.)
  - Information environment refers to a commander's battlespace that encompasses information activity affecting the operation. To envision that part of the information environment that is within their battlespace, commanders determine the information activities that affect their own operational capabilities and opposing C2 and information systems.
  - Force Projection Bases are the intermediate staging bases and power projection platforms.
  - Home Stations are the permanent locations of active and RC units. To a significant degree, events occurring at home station affect the morale and performance of deployed forces. Thus, the commander's battlespace encompasses all home station functions.

- **Battlefield Organization** is the allocation of forces in the AO by purpose. It consists of three all-encompassing categories of operations:
  - Decisive Operations are those that directly accomplish the task assigned by higher headquarters. Decisive operations conclusively determine the outcome of major operations, battles, and engagements.
  - Shaping Operations are operations at any echelon that create and preserve conditions for the success of decisive operations.
  - Sustaining Operations are operations at any echelon that enable shaping and decisive operations by providing CSS, rear area and base security, movement control, terrain management, and infrastructure development.

3-14. AOIs extend into enemy territory, to the objectives of current or planned operations. They include areas occupied by enemy forces that could jeopardize the accomplishment of the mission. AOIs also serve to focus intelligence development and IO directed at factors outside the AO that may affect the operation. (FM 3-0, para 4-79) The scheme of support includes the coordination of reconnaissance and surveillance missions and AOIs with a joint force or higher headquarters and lateral units to answer the intelligence requirements within the AOI. The G2/S2 monitors enemy activities within the AOI and provides intelligence on enemy activities that may affect the unit.

3-15. The AOIR is an area allocated to a commander in which the commander is responsible for the provision of intelligence within the means at the commander's disposal. It is an area allocated to the commander in which the commander is responsible for the collection of information concerning the threat and the environment and the analysis of that information in order to produce intelligence. Higher headquarters also ensure through intelligence handovers, collection management, and deconfliction that problems with duplication, confliction, and command and control do not occur in the AOIR. (FM 2-19.402/FM 34-80-2) They include the available ISR assets, capability of the G2/S2 section, available intelligence architecture, and METT-TC considerations. The AOIR cannot extend beyond a unit's AOI; however, it can be smaller than its AO as well as vary (expand or contract) during an operation. An example of when a unit's AOIR is smaller than its AO is when a higher headquarters ISR effort covers an area within the unit's AO.

3-16. ICLs designate the boundary between AOIRs. The G2/S2 establishes ICLs to facilitate coordination between higher, lateral, and subordinate units; coordinates with the G3/S3 to direct subordinates to track enemy units and HPTs in their areas; and hands over intelligence responsibility for areas of the battlefield. The establishment of ICLs ensures that there are no gaps in the collection effort; that all echelons are aware of the location, mission, and capabilities of other assets. The G2/S2 keeps abreast of collection activities in progress (all echelons) and battlefield developments through the ICLs. ICLs help—

- Facilitate coordination between higher, lateral, and subordinate units.

- Ensure higher, lateral, and subordinate units share information and intelligence as enemy entities and HPTs move into, within, or transition between AOIRs.
- Specify intelligence responsibility for areas of the battlefield.

ARMY CAPABILITIES

3-17. Commanders combine AC and RC ARFOR—consisting of different types of units with varying degrees of modernization—with multinational forces and civilian agencies to achieve effective and efficient unified action. A broad range of organizations makes up the institutional Army that supports the field Army. Institutional Army organizations design, staff, train, and equip the force. The institutional Army assists in effectively integrating Army capabilities. It does this through leadership and guidance regarding force structure, doctrine, modernization, and budget. (See FM 100-11.)

TASK ORGANIZATION

3-18. The Army supports JFCs by providing tailored force packages to accomplish joint missions and dominate enemies and situations on land. Trained and equipped AC and RC units comprise these force packages. Within these force packages, Army commanders organize groups of units for specific missions. They reorganize for subsequent missions when necessary. This process of allocating available assets to subordinate commanders and establishing their command and support relationships is called task organizing. A temporary grouping of forces designed to accomplish a particular mission is a task organization. The ability of ARFOR to tailor (select forces based upon a mission) and task organize (temporarily organize units to accomplish a tactical mission) gives them extraordinary agility. It allows operational and tactical level commanders to organize their units to make the best use of available resources. The ability to task organize means ARFOR can shift rapidly among offensive, defensive, stability, and support operations.

COMBINED ARMS

3-19. The fundamental basis for the organization and operations of ARFOR is combined arms. Combined arms is the synchronized or simultaneous application of several arms—such as infantry, armor, field artillery, engineers, air defense, and aviation—to achieve an effect on the enemy that is greater than if each arm was used against the enemy separately or sequentially. The ultimate goal of Army organization for operations remains success in joint and combined arms warfare. Its combined arms capability allows commanders to form Army combat, combat support (CS), and CSS forces into cohesive teams focused on common goals.

COMMAND AND SUPPORT RELATIONSHIPS

3-20. Establishing clean command and support relationships is fundamental in organizing for all operations. These relationships can achieve clear responsibilities and authorities among subordinate and supporting units. The commander designates command and support relationships within his authority to weigh the decisive operation and support his scheme of maneuver. Some forces available to a commander are given command or support relationships that limit his authority to prescribe additional relationships. Command and support relationships carry with them varying responsibilities to the subordinate unit by parent and gaining units. By knowing the inherent responsibilities of each command and support relationship, a commander may organize his force to establish clear relationships.

3-21. Command relationships establish the degree of control and responsibility commanders have for forces operating under their tactical control (TACON). When commanders establish command relationships, they determine if the command relationship includes administrative control (ADCON). Figure 3-1 shows command and support relationships.

3-22. Support relationships define the purpose, scope, and effect desired when one capability supports another. Support relationships establish specific responsibilities between supporting and supported unit. Army support relationships are direct support (DS), general support (GS), general support-reinforcing (GSR), and reinforcing.

3-23. While not an actual C2 function, technical control often affects certain intelligence operations. Technical control ensures adherence to existing policies or regulations and provides technical guidance for MI activities, particularly HUMINT, SIGINT, and CI operations. Commanders direct operations but often rely on technical expertise to plan, prepare, execute, and assess portions of the unit's collection effort. Technical control also involves translating ISR tasks into the specific parameters used to focus highly technical or legally sensitive aspects of the ISR effort. Technical control includes, but is not limited to—

- Defining, managing, or guiding the employment of specific ISR assets.
- Identifying critical technical collection criteria such as technical indicators.
- Recommending collection techniques, procedures, or assets.
- Conducting operational reviews.
- Conducting operational coordination.
- Conducting specialized training for specific MI personnel or units.

3-24. An example of technical control is the Prophet control team converting the PIR and SOR sets from the MDMP process and assigning times and anticipated enemy frequencies for subordinate Prophets to collect.

IF RELATIONSHIP IS:		Has Command Relationship with:	May Be Task Organized by:	Receives CSS from:	Assigned Position or AO By:	Provides Liaison To:	Establishes/ Maintains Communications with:	Has Priorities Established by:	Gaining Unit Can Impose Further Command or Support Relationship of:
							INHERENT RESPONSIBILITIES ARE:		
COMMAND	Attached	Gaining unit	Gaining unit	Gaining unit	Gaining unit	As required by gaining unit	Unit to which attached	Gaining unit	Attached; OPCON; TACON; GS; GSR; R; DS
	OPCON	Gaining unit	Parent unit and gaining unit; gaining unit may pass OPCON to lower HQ. Note 1	Parent unit	Gaining unit	As required by gaining unit	As required by gaining unit and parent unit	Gaining unit	OPCON; TACON; GS; GSR; R; DS
	TACON	Gaining unit	Parent unit	Parent unit	Gaining unit	As required by gaining unit	As required by gaining unit and parent unit	Gaining unit	GS; GSR; R; DS
	Assigned	Parent unit	Parent unit	Parent unit	Gaining unit	As required by parent unit	As required by parent unit	Parent unit	Not Applicable
SUPPORT	Direct Support (DS)	Parent unit	Parent unit	Parent unit	Supported unit	Supported unit	Parent unit; Supported unit	Supported unit	Note 2
	Reinforcing (R)	Parent unit	Parent unit	Parent unit	Reinforced unit	Reinforced unit	Parent unit; reinforced unit	Reinforced unit: then parent unit	Not Applicable
	General Support Reinforcing (GSR)	Parent unit	Parent unit	Parent unit	Parent unit	Reinforced unit and as required by parent unit	Reinforced unit and as required by parent unit	Parent unit; then reinforced unit	Not Applicable
	General Support (GS)	Parent unit	Parent unit	Parent unit	Parent unit	As required by parent unit	As required by parent unit	Parent unit	Not Applicable

NOTE 1. In NATO, the gaining unit may not task organize a multinational unit (see TACON).
NOTE 2. Commanders of units in DS may further assign support relationships between their subordinate units and elements of the supported unit after coordination with the supported commander.

Figure 3-1. Army Command and Support Relationships and Inherent Responsibilities.

Chapter 4
Intelligence Process in Full Spectrum Operations

THE INTELLIGENCE PROCESS

4-1. Commanders use the operations process of plan, prepare, execute, and assess to continuously design and conduct operations (see Figure 4-1). The commander cannot successfully accomplish the activities involved in the operations process without information and intelligence. The design and structure of intelligence operations support the commander's operations process by providing him with intelligence regarding the enemy, the battlefield environment, and the situation.

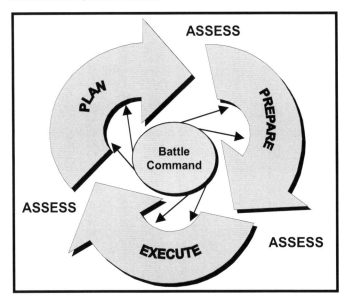

Figure 4-1. The Operations Process.

4-2. Intelligence operations consist of the functions that constitute the intelligence process: **plan, prepare, collect, process, produce,** and the three common tasks of **analyze, disseminate,** and **assess**. Just as the activities of the operations process overlap and recur as circumstances demand, so do the functions of the intelligence process. Additionally, the analyze, disseminate, and assess functions of the intelligence process occur continuously throughout the intelligence process.

FM 2-0

4-3. The operations process and the intelligence process are mutually dependent. The commander, through the operations process, provides the guidance and focus through CCIRs and PIRs that drives the intelligence process; the intelligence process provides the continuous intelligence essential to the operations process. Intelligence about the enemy, the battlefield environment, and the situation allows the commander and staff to develop a plan, seize and retain the initiative, build and maintain momentum, and exploit success (see Figure 4-2). The intelligence process is just one of the mechanisms that provides input to build the COP and facilitate the commander's situational understanding.

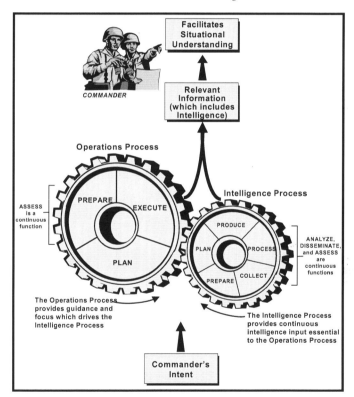

Figure 4-2. The Relationship Between the Operations and Intelligence Processes.

FM 2-0

PLAN

4-4. The planning step of the intelligence process consists of activities that include assessing the situation, envisioning a desired outcome (also known as setting the vision), identifying pertinent information and intelligence requirements, developing a strategy for ISR operations to satisfy those requirements, directing intelligence operations, and synchronizing the ISR effort. The commander's intent, planning guidance, and CCIRs (PIRs and FIRs) drive the planning of intelligence operations. Planning, managing, and coordinating these operations are continuous activities necessary to obtain information and produce intelligence essential to decisionmaking.

COORDINATE

4-5. Staff and leaders coordinate with various elements, units, and organizations to ensure the necessary resources, linguist support (see Appendix B), information, intelligence, training, and procedures are in place to facilitate effective intelligence operations.

- Coordination for Movement of ISR Assets. All ISR assets at one time or another will move through or near another unit's AO. To avoid fratricide, ISR elements must coordinate with units, G3/S3, G2/S2, and each other, as well as coordinate with the fire support officer (FSO) to establish no-fire areas and/or other control measures around ISR assets and the air defense officer (ADO) in reference to aerial ISR assets in order to establish the appropriate weapons control status.
- Coordination for Information and Intelligence. The intelligence staff must prepare and practice coordination with personnel from all MI units, non-MI units, other service components, and multinational organizations that may contribute to or facilitate the collection effort. This coordination enables the G2/S2 to share and update databases, information, and intelligence and ensures connectivity with those organizations. All units are sources of relevant information regarding the enemy and the operational environment.
- Liaison. In order to accomplish the mission, exchange information and intelligence, move through certain areas and ensure FP, it may be necessary to coordinate with many different elements, organizations, and local nationals of the country in which friendly forces are conducting operations. Local nationals include police, town officials, foreign military forces, and political and other key figures within the AO. Operations may also necessitate coordination with other US and multinational forces; for example, the International Police Task Force (IPTF), Joint Commission Observers (JCO), Organization for Security and Cooperation in Europe (OSCE), Allied Military Intelligence Battalion (AMIB), and Defense HUMINT Service (DHS).
- Movement. Coordination with the G3/S3 ensures ease of movement and safe passage of friendly forces through an area. Coordinating movement also helps avoid fratricide.

INTELLIGENCE BOS CONSIDERATIONS

4-6. The Intelligence BOS is a unified system that anticipates and satisfies intelligence needs. Commanders ensure its proper employment by clearly articulating intent, designating CCIRs (PIRs and FFIRs), and prioritizing targets. Commanders must, however, understand the limitations of the Intelligence BOS to preclude unrealistic expectations of the system. The following are Intelligence BOS limitations:

- Intelligence only reduces uncertainty on the battlefield; it does not eliminate it entirely. The commander will always have to determine the presence and degree of risk involved in conducting a particular mission.
- The Intelligence BOS is composed of finite resources and capabilities. Intelligence systems and soldiers trained in specific ISR skills are limited in any unit. Once lost to action or accident, these soldiers and systems are not easily replaceable; for some, it may not be possible to replace them during the course of the current operation. The loss of soldiers and equipment can result in the inability to detect or analyze enemy actions. The loss of qualified language-trained soldiers, especially soldiers trained in low-density languages or skills, could adversely affect intelligence operations as well.
- The Intelligence BOS cannot effectively and efficiently provide intelligence without adequate communications equipment, capacity, and connectivity. Commanders and G2/S2s must ensure communications support to intelligence has the appropriate priority.
- Commanders and G2/S2s cannot expect that higher echelons will automatically send them everything they need. While intelligence reach is a valuable tool, the push of intelligence products from higher echelons does not relieve subordinate staffs from conducting detailed analysis and focusing the efforts of higher headquarters. Nor can they expect products pushed to them to be always at the level of detail they require. Commanders and G2/S2s must focus higher echelons by clearly articulating and actively pursuing intelligence requirements. By providing higher echelons with a clear picture of the required intelligence products, commanders can also narrow the flow of intelligence and information and preclude being overwhelmed by too much information.

4-7. Commanders should be aware that intelligence collection is enabled by, and subject to, laws, regulations, and policies to ensure proper conduct of intelligence operations. While there are too many to list here specifically, categories of these legal considerations include United States Codes (USCs), Executive Orders, National Security Council Intelligence Directives (NCSIDs), Army Regulations, United States Signal Intelligence Directives (USSIDs), SOFAs, ROE, and other international laws and directives.

PREPARE

4-8. The prepare step includes those staff and leader activities which take place upon receiving the OPORD, OPLAN, WARNO, or commander's intent

to improve the unit's ability to execute tasks or missions and survive on the battlefield. These activities include—
- Effecting necessary coordination in accordance with the OPORD, METT-TC, unit SOP.
- Establishing and testing the intelligence architecture. This activity includes complex and technical issues like hardware, software, communications, COMSEC materials, network classification, technicians, database access, liaison officers (LNOs), training, funding, and TTP.
- Establishing an intelligence team attitude. This activity includes knowing different unit's and organization's capabilities, training the necessary collective skills, establishing effective relationships with different units and organizations, developing mutual battle rhythms and TTP, and leveraging the right architectures and collaboration tools.
- Coordinating effective analytic collaboration. Effective analytic collaboration is necessary to maximize the complementary analytic capabilities of different units and organizations that produce intelligence within the same theater of operations. Coordinating this collaboration is an effort-intensive activity that requires careful mutual planning, division of labor, defined responsibilities, and procedures for adapting to changing circumstances as they develop.
- Establishing reporting procedures.
- Conducting IPB.
- Producing Intelligence Estimates.
- Presenting briefings.
- Ensuring staff and personnel are trained. If personnel are not adequately trained at this point, they must be trained or the leader must evaluate the risk they bring to the operation.
- Planning refinement, brief-backs, SOP reviews, rehearsals, and coordinating with various elements and organizations.
- Establishing other troop-leading procedures (TLPs) or coordination, as necessary, in accordance with METT-TC factors.

G2/S2 PREPARATION ACTIVITIES

4-9. The G2/S2 takes numerous steps before mission execution to ensure intelligence operations run smoothly and effectively. These steps include, but are not limited to, the following:
- Conduct rehearsals.
- Conduct communication rehearsals.
- Review and update available databases and IPB products.
- Review applicable SOPs, Army Regulations, DA Pamphlets, Field Manuals, and ROE for guidance in conducting intelligence operations.
- Plan and practice actions supporting likely contingencies, or the branches or sequels to an operation.
- Ensure coordination measures are still in effect.
- Ensure training (individual and collective).

- Verify communications protocols with theater and higher headquarters and subordinate and lateral units.
- Update intelligence databases.
- Update the forces with the most recent intelligence on the AO immediately before mission execution.

VERIFICATION

4-10. Coordination for or requesting provisions or services is only the first step in acquiring them. It is crucial that staff and leaders check to verify that procedures, personnel, equipment, and services are in place and ready for mission execution.

REHEARSALS

4-11. Rehearsals help units prepare for operations by either verifying that provisions and procedures are in place and functioning or identifying inadequacies, which staff and leaders must remedy. They allow participants in an operation to become familiar with and to translate the plan into specific actions that orient them to their environment and other units when executing the mission. They also imprint a mental picture of the sequence of key actions within the operation and provide a forum for subordinate and supporting leaders and units to coordinate. (FM 6-0)

REPORTING PROCEDURES

4-12. The timely and accurate reporting of CCIRs (PIRs and FFIRs) and IRs is key to successful operations. All assets should know when, how often, and what format to use when reporting. The G2/S2 must verify the frequencies, alternate frequencies, and reactions during jamming, as well as the LTIOV for specific information to be reported. Unit SOPs provide the proper reporting procedures.

4-13. The G2/S2 coordinates with the unit staff, subordinate and lateral commands, and higher echelon units to ensure that specific reporting assets, personnel, equipment (especially communications), and procedures are in place. The G2/S2 requests or establishes the appropriate message addresses, routing indicators, mailing addresses, and special security office (SSO) security accreditation for units.

COMMUNICATIONS

4-14. Staff and leaders must work closely with the G6/S6 or signal officer (SIGO) to coordinate for the required communication links. The unit may require classified and unclassified network connections for their equipment. If elements of the unit will be working outside the range of the unit's communications systems, then it is necessary to coordinate for global or extended range communications. Leaders must obtain the required type and amount of communications equipment and related components as well as the latest fills and frequencies. They must possess and be familiar with all the instructions, passwords, policies, regulations, and directives conducive to OPSEC. They must also ensure soldiers are trained in the use and procedures involved in operating communications equipment.

SITUATION UPDATES

4-15. Each staff section and element conducts activities to maximize the operational effectiveness of the force. Coordination and preparation are just as important, if not more important, as developing the plan. Staff preparation includes assembling and continuously updating estimates. For example, continuous IPB provides accurate situational updates for commanders.

INTELLIGENCE HANDOFF

4-16. A well-executed intelligence hand-off will ensure a smooth and seamless transition between units. It is important that the incoming unit becomes familiar with the operation as soon as possible to avoid compromising the intelligence production and flow of the mission. The following are points to consider during a mission hand-off:

- Briefings and reports (learn what briefings are required and when as well as report formats and requirements).
- Past, present, and planned activities within the AOI.
- Established SOPs (know procedures for reporting; intelligence contingency funds [ICFs] and incentive use if applicable; emplacement and use of ISR equipment).
- Key personalities (introductions are required; establish rapport and a good working relationship with all key personalities).
  - Key personnel on the base or camp (their responsibilities; how to contact them).
  - Key personnel in other US and multinational service components (coordinate for exchange of information and intelligence).
  - Key personalities from surrounding towns (local figures).
  - Key national level political and military figures.
- Supporting units (know where to go for provisions, information, or assistance and POCs within those organizations).
- Current attitudes (understand current attitudes and perspectives of the local populace).
- Equipment operation and idiosyncrasies (equipment may run on different applications; personnel may need to train on specific equipment and procedures).
- Area familiarization (identify NAIs, key terrain, minefields, and boundaries; know camp locations, routes and route names, checkpoints, towns, and troubled resettlement areas).

RULES OF ENGAGEMENT

4-17. Although ROE training was presented during the plan function of the intelligence process, leaders at all levels can take the opportunity during the prepare function to ensure their subordinates completely understand the ROE. It is also during this function that commanders may need to consider exceptions to, or modifications of, the ROE to facilitate HUMINT and CI collection or to enable the placement of ISR assets.

COLLECT

4-18. Recent ISR doctrine necessitates that the entire staff, especially the G3/S3 and G2/S2, must change their reconnaissance and surveillance mindset to conducting ISR. The staff must carefully focus ISR on the CCIR (PIR and FFIR) but also enable the quick retasking of units and assets as the situation changes. This doctrinal requirement ensures that the enemy situation not just our OPLAN "drives" ISR operations. Well-developed procedures and carefully planned flexibility to support emerging targets, changing requirements, and the need to support combat assessment is critical. The G2/S2 and G3/S3 play a critical role in this challenging task that is sometimes referred to as "fighting ISR" because it is so staff intensive during planning and execution (it is an operation within the operation). Elements of all units on the battlefield obtain information and data about enemy forces, activities, facilities, and resources as well as information concerning the environmental and geographical characteristics of a particular area.

ISR TASKS

4-19. ISR tasks are the actions of the intelligence collection effort. ISR tasks consists of three categories:
- Intelligence.
- Surveillance.
- Reconnaissance.

4-20. Intelligence tasks are included in Annex B of the OPORD under Scheme of Intelligence. They include the following:
- **Intelligence Production.** Intelligence production includes analyzing information and intelligence and presenting intelligence products, conclusions, or projections regarding the OE and enemy forces in a format that enables the commander to achieve situational understanding.
- **Request for Information.** Submitting an RFI to the next higher or lateral echelon is the normal procedure for obtaining intelligence information not available through the use of available ISR assets. Users enter RFIs into an RFI management system where every other user of that system can see it. Hence, an echelon several echelons above the actual requester becomes aware of the request and may be able to answer it. A G2/S2 who receives an RFI from a subordinate element may use intelligence reach to answer RFIs.
- **Intelligence Reach.** Intelligence reach allows the commander to access the resources of national, joint, foreign, and other military organizations and units. Requestors can acquire information through push and pull of information, databases, homepages, collaborative tools, and broadcast services. (See Chapter 2 for more information on intelligence reach.)

4-21. For information on reconnaissance or surveillance tasks, refer to FM 7-15 and FM 3-55.

SPECIAL RECONNAISSANCE

4-22. Special reconnaissance (SR) is the complementing of national and theater intelligence collection assets and systems by obtaining specific, well defined, and time-sensitive information of strategic or operational significance. It may complement other collection methods where there are constraints of weather, terrain, hostile countermeasures, and/or other systems availability. SR places US or US-controlled personnel conducting direct observation in hostile, denied, or politically sensitive territory when authorized. SOF may conduct these missions unilaterally or in support of conventional operations. (See JP 3-05 and FM 101-5-1.)

4-23. Army Special Operations Forces (ARSOF) elements conduct SR missions to obtain information not available through other means. SR operations encompass a broad range of collection activities to include reconnaissance, surveillance, and TA. SR complements national and theater collection systems that are more vulnerable to weather, terrain masking, and hostile countermeasures. SR missions provide intelligence or information that is often not available through other means. Typical SR missions include—

- TA and surveillance of hostile C2 systems, troop concentrations, deep-strike weapons, lines of communication (LOCs), WMD systems, and other targets.
- Location and surveillance of hostage, prisoner of war, or political prisoner detention facilities.
- Post-strike reconnaissance for BDA.
- Meteorologic, geographic, or hydrographic reconnaissance to support specific air, land, or sea operations.

4-24. For more information on special reconnaissance, see FM 3-05.102.

REPORTING

4-25. The most critical information collected is worthless if not reported in a timely manner. Collectors may report information via verbal, written, graphic, or electronic means. Unit SOPs must clearly state the transmission means of different types of reports (for example, sent by voice frequency modulated [FM] radios or by automated means). In general, the transmission of reports for enemy contact and actions, CCIRs, exceptional information, and NBC reports is by voice FM, and then followed up with automated reports. Commanders and staffs must remember that timely reporting, especially of enemy activity, is critical in fast-moving operations. Collectors must report accurate information as quickly as possible. Commanders and staff must not delay reports for the sole purpose of editing and ensuring the correct format. This is particularly true for reporting information or intelligence that answers the PIR.

TIME-SENSITIVE REPORTING

4-26. Intelligence and time-sensitive combat information that affects the current operation is disseminated immediately upon recognition. Combat information is unevaluated data, gathered by or provided directly to the tactical commander which, due to its highly perishable nature or the criticality of the situation, cannot be processed into tactical intelligence in time to satisfy the user's tactical intelligence requirements. Thus combat information is provided directly to the tactical commander (see JP 1-02). The routing of combat information proceeds immediately in two directions: directly to the commander and through routine reporting channels, which include intelligence analysis and production elements.

4-27. Time-sensitive information usually includes reports concerning enemy contact and actions and CCIRs.

PROCESS

4-28. The process function converts relevant information into a form suitable for analysis, production, or immediate use by the commander. Processing also includes sorting through large amounts of collected information and intelligence (multidiscipline reports from the unit's ISR assets, lateral and higher echelon units and organizations, and non-MI elements in the battlespace). Processing identifies and exploits that information which is pertinent to the commander's intelligence requirements and facilitates situational understanding. Examples of processing include developing film, enhancing imagery, translating a document from a foreign language, converting electronic data into a standardized report that can be analyzed by a system operator, and correlating dissimilar or jumbled information by assembling like elements before the information is forwarded for analysis.

4-29. Often collection assets must collect and process their data prior to disseminating it. MI systems have their own reporting and processing systems, the details of which are in the appropriate MI system manuals and technical manuals. Some collection assets, particularly air reconnaissance and ground scouts, can report relevant information that is immediately usable by the tactical commander (for example, for targeting purposes). However, the personnel in the reporting chain still process these reports by evaluating their relevancy and accuracy. In many cases, the output of a collection asset is data, or information of limited immediate use to a commander. Also, in certain situations ROE dictate a requirement for target confirmation by other sources.

4-30. The intelligence staff processes information collected by the unit's assets as well as that received from higher echelons. The intelligence staff processes many types of information and data from intelligence reach, unmanned aerial vehicle (UAV) imagery, radar imagery, mobile target indicators (MTIs), and HUMINT and SIGINT reports.

PRODUCE

4-31. In the production step, the G2/S2 integrates evaluated, analyzed, and interpreted information from single or multiple sources and disciplines into finished intelligence products. Like collection operations, the G2/S2 must ensure the unit's information processing and intelligence production are prioritized and synchronized to support answering the CCIRs (PIRs and FFIRs).

4-32. Intelligence products must be timely, relevant, accurate, predictive, and usable. The accuracy and detail of every intelligence product has a direct effect on how well the unit plans and prepares for operations. However, the G2/S2 and unit must use intelligence (no matter what form the intelligence is in) that meets the requirements but might not be as detailed or refined as possible or in a better form. A good answer on time is better than a more refined answer that is late.

4-33. The G2/S2 produces intelligence for the commander as part of a collaborative process. The commander drives the G2/S2's intelligence production effort by establishing intelligence and IRs with clearly defined goals and criteria. Differing unit missions, environments, and situations impose numerous and varied production requirements on the G2/S2 and his staff.

4-34. The G2/S2 must employ collaborative analysis techniques and procedures that leverage intelligence production capability of higher and subordinate echelons to meet these requirements. Proficiency in these techniques and procedures enables the G2/S2 to answer the commander's and staff's requirements regardless of the mission, environment, and situation. The G2/S2 and staff intelligence products enable the commander to—

- Plan operations and employ maneuver forces effectively.
- Recognize potential COAs.
- Employ effective tactics and techniques.
- Take appropriate security measures.
- Focus ISR.

COMMON INTELLIGENCE PROCESS TASKS

ANALYZE

4-35. Analysis occurs at various stages throughout the intelligence process. Personnel conducting intelligence operations at all levels analyze intelligence, information, and problems to produce intelligence, solve problems and, most importantly, answer the PIRs. Leaders at all levels conduct analysis to assist in making many types of decisions. An example is a HUMINT collector analyzing an intelligence requirement in order to determine the best possible collection strategy to use against a specific source.

4-36. Analysis in RM is critical to ensuring the IRs receive the appropriate priority for collection. The intelligence staff analyzes each requirement to determine its feasibility, whether or not it supports the commander's intent, and to determine the best method of satisfying the IRs. The staff also analyzes collected information to determine if it satisfies requirements.

4-37. During the produce function, the intelligence staff analyzes information from multiple sources to develop all-source intelligence products. The intelligence staff analyzes information and intelligence to ensure the focus, prioritization, and synchronization of the unit's intelligence production is in accordance with the PIRs.

4-38. In situation development, the intelligence staff analyzes information to determine its significance relative to predicted ECOAs and the CCIRs (PIRs and FFIRs). Through predictive analysis, the staff attempts to identify enemy activity or trends that represent opportunities or risks to the friendly force.

They use the indicators developed for each ECOA and CCIRs (PIRs and FFIRs) during the MDMP as the basis for their analysis and conclusions.

DISSEMINATE

4-39. Successful operations at the tactical and operational levels require an increased ability to synchronize fires, have faster access to intelligence, and enhance situational understanding and effective FP. Timely and accurate dissemination of intelligence is key to the success of these and other operations. Commanders must receive combat information and intelligence products in time and in an appropriate format to support decisionmaking. Additionally, sharing the most current all-source information and intelligence at all echelons is essential for commanders to maintain situational understanding.

4-40. To achieve this, it is imperative that the commander and staff establish and support a seamless intelligence architecture—including an effective dissemination plan—across all echelons to ensure information and intelligence flow in a timely manner to all those who need them. Intelligence and communications systems continue to evolve in their sophistication, application of technology, and accessibility to the commander. Their increasing capabilities also create an unprecedented volume of information available to commanders at all echelons. Finally, the commander and staff must have a basic understanding of these systems and how they contribute to the Intelligence BOS. A dissemination plan can be a separate product, or integrated into existing products such as the ISR synchronization plan or ISM, the decision support template (DST), or decision support matrix (DSM).

Dissemination Procedures

4-41. The G2/S2 and intelligence personnel at all levels assess the dissemination of intelligence and intelligence products.

4-42. Disseminating intelligence simultaneously to multiple recipients is one of the most effective, efficient, and timely methods. This can be accomplished through various means; for example, push, broadcast. However, within the current tactical intelligence architecture, reports and other intelligence products move along specific channels. The staff helps streamline information distribution within these channels by ensuring dissemination of the right information in a timely manner to the right person or element. There are three channels through which commanders and their staffs communicate: command, staff, and technical.

4-43. **Command Channel.** The command channel is the direct chain-of-command link that commanders, or authorized staff officers, use for command-related activities. Command channels include command radio nets (CRNs), video teleconferences (VTC), and the Maneuver Control System (MCS).

4-44. **Staff Channel.** The staff channel is the staff-to-staff link within and between headquarters. The staff uses the staff channel for control-related activities. Through the staff channel, the staff coordinates and transmits intelligence, controlling instructions, planning information, provides early warning information, and other information to support C2. Examples of staff channels include the operations and intelligence radio net, telephone, the staff huddle, VTC, and the BOS-specific components of the Army Battle Command System (ABCS).

4-45. **Technical Channel.** Staffs typically use technical channels to control specific combat, CS, and CSS activities. These activities include fire direction and the technical support and sensitive compartmented information (SCI) reporting channels of intelligence and ISR operations. The SIGINT tasking and reporting radio net, intelligence broadcast communications, and the wide area networks (WANs) supporting single intelligence discipline collection, processing, and production are examples of technical channels.

Presentation Techniques and Procedures

4-46. The staff's objective in presenting information is to provide the commander with relevant information. Table 4-1 lists the three general methods that the staff uses to present information and meet its information objective. Systems within the ABCS contain standard report formats, maps, and mapping tools that assist the staff in presenting information in written, verbal, and graphic form. Audio and video systems such as large format displays and teleconferencing systems enable the staff to use a combination of the methods in multimedia presentations.

Table 4-1. Presentation Methods and Products.

Method	Products
Written Narrative	Reports, Estimates, and Studies
Verbal Narrative	Briefing (information, decision, mission, and staff)
Graphic	Charts, Overlays, and Electronic Displays

Intelligence Communications Architecture

4-47. The intelligence communications architecture transmits intelligence and information to and from various ISR elements, units, and agencies by means of automation and communication systems. With the continued development of sensors, processors, and communications systems, it is increasingly important to understand the requirements of establishing an effective communications architecture. The G2/S2 must identify the Intelligence BOS specific requirements of the unit's overall communications architecture. Refer to FM 2-33.5 for more information on intelligence communications reach. The following are some (but not all) of the questions which the staff must answer in order to establish the intelligence communications architecture:

- Where are the unit's collectors?
- What and where are the unit's processors?
- Where are the unit's intelligence production elements?
- Where are the unit's decisionmakers?
- How does the unit disseminate information from its producers to its decisionmakers and/or consumers?

- Are the systems which the unit's collectors, producers, processors, and consumers use compatible with each other? If not, what is the plan to overcome this challenge?
- How can the unit access databases and information from higher and other agencies?

ASSESS

4-48. Assessment is the continuous monitoring—throughout planning, preparation, and execution—of the current situation and progress of an operation, and the evaluation of it against criteria of success to make decisions and adjustments. Assessment plays an integral role in all aspects of the intelligence process. Assessing the situation and available information begins upon receipt of the mission and continues throughout the intelligence process. The continual assessment of intelligence operations and ISR assets, available information and intelligence, the various aspects of the battlefield environment, and the situation are critical to—

- Ensure the CCIRs (PIRs and FFIRs) are answered.
- Ensure intelligence requirements are met.
- Redirect collection assets to support changing requirements.
- Ensure operations run effectively and efficiently.
- Ensure proper use of information and intelligence.
- Identify enemy efforts at deception and denial.

4-49. During planning, the intelligence staff conducts a quick initial assessment of the unit's intelligence posture and holdings, status of intelligence estimates, and any other available intelligence products. From this assessment the commander issues his initial guidance and a WARNO.

4-50. While the majority of the unit is engaged in preparation, the ISR effort should already have begun. It is during this period when the prepare and execute activities of the operations process overlap, that the G2/S2 assesses the current situation as well as the progress of ISR operations.

4-51. During execution the intelligence staff continues assessing the effectiveness of the ISR effort while at the same time assessing the results and products derived from the collection effort. The critical aspects of assessment at this point include determining whether the PIRs have been answered, will be answered with the current ISR operations, or which ISR operations to adjust in order to answer the CCIRs (PIRs and FFIRs). This type of assessment requires sound judgment and a thorough knowledge of friendly military operations, characteristics of the AO and AOI, and the threat situation, doctrine, patterns, and projected future COAs.

PART THREE
Military Intelligence Disciplines

Part Three provides a more detailed explanation of the intelligence disciplines introduced in Part One of this manual.

Chapter 5 defines and discusses the roles and fundamentals of the all-source intelligence discipline.

Chapter 6 defines and discusses the roles, fundamentals, and generic organization of the HUMINT discipline.

Chapter 7 defines and discusses the roles, fundamentals, and generic organization of the IMINT discipline.

Chapter 8 defines and discusses the roles, fundamentals, and generic organization of the SIGINT discipline.

Chapter 9 defines and discusses the roles, fundamentals, and generic organization of the MASINT intelligence discipline.

Chapter 10 defines and discusses the roles, fundamentals, and generic organization of the TECHINT discipline.

Chapter 11 defines and discusses the roles, fundamentals, and generic organization of the CI discipline.

Chapter 5
All-Source Intelligence

DEFINITION

5-1. All-source intelligence is defined as the intelligence products, organizations, and activities that incorporate all sources of information and intelligence, including open-source information, in the production of intelligence. All-source intelligence is a separate intelligence discipline, as well as the name of the function used to produce intelligence from multiple intelligence or information sources.

ROLE

5-2. The operational environment provides an ever-growing volume of data and information available from numerous sources, from which the commander can use to achieve situational understanding. His situational understanding enables him to make decisions in order to influence the outcome of the operation; prioritize and allocate resources; assess and take risks; and understand the needs of the higher and subordinate commanders. The commander depends upon a skilled G2/S2 working within his intent to support his ISR effort and provide all-source intelligence analysis conclusions and projections of future conditions or events.

FUNDAMENTALS

5-3. All-source intelligence production satisfies intelligence requirements. It provides an overall picture of the adversary and the battlespace. It reduces the possibility of error, bias, and misinformation through the use of multiple sources of information and intelligence.

PLAN

5-4. The utilization of all-source intelligence facilitates the development of accurate and concise contingency-specific plans, orders, and intelligence products. Additionally, the G2/S2 must retrieve, update, or develop any required intelligence databases. The most important all-source products are the modified combined obstacle overlay (MCOO), event templates, ECOA sketches, and the HPT list provided during initial IPB in support of the plan function. Through all-source intelligence, the commander can make informed decisions about COAs and adversary capabilities.

5-5. The intelligence staff coordinates its efforts with other elements such as the engineer terrain team, the unit surgeon, the Air Force weather team, and other assets or elements that can support the analytical effort.

5-6. IPB plays a significant role in the planning phase of the intelligence process. It is a systematic, continuous process of analyzing the threat and environment and is designed to support staff estimates and the MDMP. The IPB is led by the G2/S2, with participation by the entire staff. IPB allows the commander and staff to make informed decisions, develop COAs, and focus ISR efforts where they are most needed. IPB continuously assists the commander and staff in focusing ISR assets on the appropriate targets at critical points in time and space.

5-7. The time available for the IPB process may not permit the luxury of conducting each step in detail. Overcoming time limitations requires the commander to identify those parts of IPB that are most important to the commander in planning and executing his mission. Applying the specific steps or degree of detail performed varies according to METT-TC. There are four steps in the IPB process: define the battlefield environment, describe the battlefield's effects, evaluate the threat, and determine ECOAs.

Define the Battlefield Environment

5-8. In this step, the G2/S2—
- Identifies characteristics of the battlefield that will influence friendly and threat operations including terrain (mobility), weather, hydrological data, infrastructure, and civilian demographics. Produces accurate, timely, and predictive IPB products that depict the aspects of the battlefield.

- Identifies the limits of the command's AO and AOI.
- Identifies gaps in current intelligence holdings, identifies IRs, and recommends CCIRs (PIRs and FFIRs).

5-9. Defining the battlefield environment includes identifying enemy forces (their location, mobility, general capabilities, and weapon ranges) and all other aspects of the environment that could have an effect on the unit's ability to accomplish the mission. Depending on the situation, these considerations may also include—

- Geography, terrain, and weather of the area.
- Information environment to include but not limited to computer and communications systems and capabilities, data acquisition systems and capabilities, media access and distribution, areas prone to electromagnetic interferences, and systems to generate electromagnetic interference.
- Population demographics (ethnic groups, religious sects, age distribution, health status, income groups).
- Medical threat to include but not limited to endemic and epidemic diseases, occupational and environmental health hazards, and poisonous and toxic plants and animals.
- Political or socio-economic factors, including factions, clans, and gangs.
- Infrastructures, such as transportation or telecommunications, and critical decisionmaking infrastructures and supporting information systems.
- ROE or legal restrictions, such as international treaties or agreements.

Describe the Battlefield's Effects

5-10. This IPB step deals with the effects of the battlefield environment on the current operations and potential enemy and friendly COAs. It begins with the assessment of existing and projected conditions of the battlefield environment, which the staff accomplishes through terrain analysis, weather analysis, and analysis of other characteristics of the battlefield. To conduct terrain analysis, the staff uses maps, reconnaissance, and other specialized terrain products (maps, overlays, databases, software). These products address such factors as wet or dry cross-country mobility, transportation systems (road and bridge information), vegetation type and distribution, surface drainage and configuration, surface materials (soils), ground water, manmade structures, and obstacles. The results of evaluating the terrain's effects should be expressed by identifying areas of the battlefield that influence each COA.

Evaluate the Threat

5-11. Evaluating the threat involves determining the threat force capabilities and the doctrinal principles and TTP that threat forces prefer to employ. The result of this evaluation produces a threat model, which portrays how threat forces normally execute operations and how they have reacted to similar situations in the past. The threat model includes an evaluation of the threat's strengths, weaknesses, and vulnerabilities, including an evaluation of typical HVTs.

Determine ECOAs

5-12. Determining ECOAs involves the identification and development of likely ECOAs that will influence accomplishing the friendly mission. Developing ECOAs is a form of predictive intelligence analysis and production. The procedures for this step include—

- Identifying the enemy's likely objectives and predicting the desired end-state.
- Identifying the full set of COAs available to the enemy.
- Predicting the enemy's most likely and most dangerous COAs.
- Evaluating and prioritizing each COA.
- Developing each COA in the amount of detail that time permits.
- Identifying initial ISR and collection requirements.

5-13. After conducting the initial IPB, the staff, primarily the G2/S2, identifies gaps in the available intelligence, develops the initial PIRs and IRs, and develops the initial input to the ISR plan based on the commander's guidance.

5-14. Indicators are the basis for situation development. Indicators are activities that will confirm or deny the event specified in an intelligence requirement. They are any positive or negative evidence of enemy activity or characteristic of the AO that points toward enemy capabilities, vulnerabilities, or intentions. Individual indicators cannot stand alone. Each indicator is integrated with other factors and indicators before patterns are detected and enemy intentions established. Indicators are developed by the analysts in the G2/S2 section. The event matrix shows the threat activities, or indicators, to look for in each NAI, and the timelines during which each NAI should be active. All indicators are developed to answer the commander's PIRs and IRs. The analyst uses indicators to correlate particular events or activities with probable ECOAs and to determine what events or activities must occur for an enemy to follow a particular COA. The ability to read indicators (including recognition of enemy deception indicators) contributes to the success of friendly operations. The analyst integrates information from all sources to confirm indicators of enemy activities. As indicators are detected and confirmed, PIRs are answered.

PREPARE

Activities

5-15. All-source activities during the prepare function include—

- Conducting rehearsals.
- Conducting communications rehearsals and verifiying communications protocols with higher, lateral, and subordinate units.
- Planning and practicing actions that support likely contingencies, branches, or sequels.
- Reviewing and updating available databases and IPB products.
- Ensuring control and coordination measures are still in effect.

FM 2-0

- Updating the force with the most recent intelligence before mission execution.

Intelligence Estimates

5-16. The intelligence estimate is a continuous process that is the product of all actions the intelligence staff performs throughout the MDMP. The intelligence estimate provides a timely and accurate evaluation of the enemy and the AO (and often the AOIR) at a given time. It provides the background the G2/S2 uses to portray enemy actions during COA analysis.

5-17. The G2/S2 must clearly understand the weather and terrain effects and the ability to visualize the battlespace before producing the intelligence estimate. This understanding facilitates accurate assessments and projections regarding the enemy: enemy situation (including strengths and weaknesses), enemy capabilities and an analysis of those capabilities (COAs available to the enemy), and conclusions drawn from that analysis. The intelligence estimate's conclusion identifies the enemy's most likely COA and most dangerous COA. ECOAs must include a sketch of the COA with a task and purpose for the enemy's actions. Any gaps in the intelligence estimate are identified as information requirements.

5-18. The intelligence estimate may be written or oral. At the tactical level, especially during operations and exercises, intelligence estimates are usually delivered orally, supported by graphic displays and other decision support tools. During contingency planning, especially at corps level and above, estimates are usually written. During deliberate planning at joint headquarters, estimates are always written (see JP 5-00.2). However, the intelligence estimate should always be prepared as thoroughly as time and circumstances permit. A comprehensive intelligence estimate considers both the tangible (quantifiable) and the intangible aspects of the enemy's operations. It translates enemy strengths, weapon systems, training, morale, and leadership into combat capabilities and projections of future enemy actions.

5-19. Different sections of the intelligence estimate receive more emphasis during different activities within the operations process.

- During planning, the most important decision the commander makes is selecting a COA on which to base the plan. Thus, the sections of the intelligence estimate that focus on the commander's selecting the most appropriate COA are the most important at that time.
- During preparation, the intelligence estimate must focus on the commander's decisions that affect the ability of the unit to execute the upcoming operation. The intelligence estimate—and functions—that supports these decisions must focus on answering the CCIRs (PIRs and FFIRs) and guiding the ISR effort.
- During execution, the intelligence estimate focus is to support the anticipated command decisions. The most important action of the intelligence estimate—and the ISR effort— is to answer the PIR. However, it is also during the execution phase that the intelligence estimate must look ahead: anticipating branches or sequels to the

FM 2-0

current operation, transition to other operations, changes in the ECOAs, and required adjustments to the ISR effort.

5-20. Refer to FM 2-01 for examples of intelligence estimates.

COLLECT

5-21. The collect intelligence process function within the all-source intelligence discipline is limited to gathering the necessary information, intelligence, and intelligence products required to perform analysis. Thus, all-source intelligence depends upon the other intelligence disciplines to perform collection. This is also the primary reason why OSINT is defined as a source of information and intelligence and not a separate intelligence discipline.

PROCESS

5-22. The process function converts relevant information into a form suitable for analysis, production, or immediate use by the commander. Processing also includes sorting through large amounts of collected information and intelligence to identify and exploit the information which is pertinent to the commander's intelligence requirements and facilitates situational understanding.

PRODUCE

5-23. All-source intelligence analysis and production—

- Drives collection to answer the PIR.
- Provides the enemy situation.
- Provides INTSUMs and other intelligence reports.
- Supports situational understanding.
- Provides predictive estimates of enemy actions; specifically, ECOAs.
- Provides continuously updated IPB.
- Provides all-source target packages (or folders).

5-24. As previously stated, the intelligence estimate is a continuous process. The commander and staff constantly collect, process, store, display, and disseminate information. Staff members update their estimates as they receive new information, such as—

- When they recognize new facts.
- When they replace assumptions with facts or find their assumptions invalid.
- When they receive changes to the mission or when changes are indicated.

5-25. Technological advances and NRT information allow estimates to be continuously updated; the running estimate and its intelligence component, the intelligence running estimate, exemplify this.

Running Estimate

5-26. Running estimates are continuously updated estimates based on new information as the operation proceeds. They serve as a staff technique to support the commander's visualization and decisionmaking, as well as the staff's tool for assessing during preparation and execution. In the running estimate, staff officers continuously update their conclusions and recommendations as they evaluate the impact of new facts. Current doctrine emphasizes the COP as the primary tool that provides the current situation and, when merged with the running estimates, facilitates commanders achieving situational understanding. All staff sections (BOSs) provide their respective input to the COP. The COP as used today, combined with the running estimate, is predictive and enhances our current ability to collect, process, store, display, and disseminate information.

5-27. The portion of the COP that depicts the enemy situation is currently limited to displaying the locations and dispositions of enemy and threat forces in a relatively static manner, sometimes referred to as snapshots in time. The enemy situation portion of the COP requires analysis to provide the required level of detail. The COP and running estimates sufficiently provide effective support for battle command of knowledge-based organizations.

5-28. The main differentiation between the running estimate and the old staff estimates is the emphasis on not only continuously updating the facts of the estimate but also continuously updating the conclusions and recommendations while including projections of future conditions of the entire battlespace. While the COP is primarily a display of current intelligence and information, the running estimate requires the merging of the staff's cognitive processes with automation applications. The primary focus of the staff's cognitive process is to present predictive or anticipatory intelligence in support of the commander's decisionmaking or situational understanding. See Figure 5-1.

5-29. The running estimate is a product of the entire battle staff. Just as all BOSs contribute to the COP, they will also contribute their portion of the running estimate. The running estimate integrates the running estimates from each BOS. The Intelligence BOS input to the running estimate is the intelligence running estimate.

Intelligence Running Estimate

5-30. The intelligence running estimate is a continuous flow and presentation of relevant information and predictive intelligence that, when combined with the other staff running estimates, enables the decisionmaker's visualization and situational understanding of the AOI in order to achieve information superiority. The intelligence running estimate requires constant verification to support situational understanding of the current situation as well as predictive assessments for future operations.

5-31. While it is possible to employ the concept of the intelligence running estimate today, the true seamless and continuous update feature will be achieved with technological enablers that are not yet present in the force. In the future, technology should allow the information from the running estimate to become implanted as a part of the COP. However, there will always be a place for some separate context, orally or written, separate from the COP display to add fidelity and assist in the commander's visualization and decisionmaking. Additional training must be implemented not just

FM 2-0

within MI but also throughout the force in order to institute the running estimate concept.

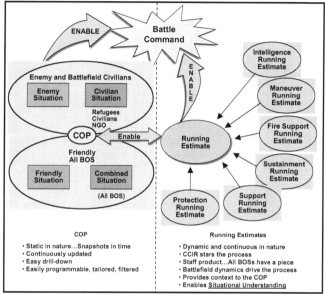

Figure 5-1. Example of the COP and a Running Estimate.

ANALYZE

5-32. Analysis is key in converting combat information and all intelligence from each discipline into all-source intelligence products and targeting information. From the multiple sources of information received, the staff analyzes and identifies critical information; determines the relevancy and accuracy of the information; and reaches conclusions about the information. These conclusions are either immediately disseminated or are used to form the basis of intelligence analysis products.

DISSEMINATE

5-33. The intelligence staff extracts and disseminates pertinent intelligence from products developed as a result of all-source analysis. The intelligence staff—

- Rapidly disseminates time-sensitive, all-source analysis intelligence and intelligence products to higher, lateral, and subordinate commands in order to keep all commanders abreast of current developments in the situation and battlespace in accordance with unit SOPs.

FM 2-0

- Disseminates all-source analysis intelligence to other G2/S2 intelligence activities for additional processing or detailed analysis and exploitation in accordance with unit SOPs.

5-34. For intelligence reach operations all-source intelligence products are available and disseminated in a variety of forms. It is incumbent on the requestor to ensure that the all-source product can be transmitted over the available communications systems. This includes verifying the appropriate security level of the communication system.

ASSESS

5-35. The commander, the intelligence staff, and the intelligence consumers assess the analysis, production, and dissemination of all-source intelligence. The staff identifies intelligence gaps, emerging operational requirements, duplication of effort, or new targets or threats and redirects all-source activities as appropriate to meet intelligence requirements. Throughout the assessment process, the G2/S2, the intelligence staff, and intelligence users should be alert for evidence of possible enemy denial and deception efforts. The staff assesses intelligence and information for—

- Consistency with the current situation and threat trends.
- Accuracy and confirmation by other intelligence sources or disciplines.
- Source reliability and credibility.
- Pertinence to PIRs, intelligence requirements, RFIs, and other intelligence tasks or requests.

Chapter 6

Human Intelligence

DEFINITION

6-1. HUMINT is the collection by a trained HUMINT Collector of foreign information from people and multimedia to identify elements, intentions, composition, strength, dispositions, tactics, equipment, personnel, and capabilities. It uses human sources as a tool and a variety of collection methods, both passively and actively, to gather information to satisfy the commander's intelligence requirements and cross-cue other intelligence disciplines.

ROLE

6-2. The role of HUMINT Collectors is to gather foreign information from people and multiple media sources to identify adversary elements, intentions, composition, strength, dispositions, tactics, equipment, personnel, and capabilities. It uses human sources and a variety of collection methods to gather information to satisfy the commander's intelligence requirements, and cross-cue other intelligence disciplines.

HUMINT FUNCTIONS

6-3. HUMINT functions are interrelated, mutually supporting, and can be derived from one another. No single function or technical capability can provide a full understanding of our adversaries. HUMINT functions are defined below and shown in Figure 6-1.

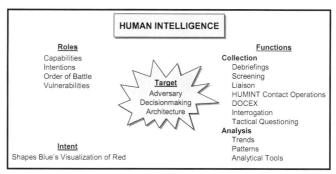

Figure 6-1. HUMINT Roles and Functions.

COLLECTION

6-4. HUMINT collection methods and operations include—

- **Debriefings.** Debriefings are the systematic questioning of individuals to procure information to answer specific collection requirements by direct and indirect questioning techniques. Sources for debriefings are categorized as friendly forces, US and non-US civilians to include refugees, displaced persons, and local inhabitants.
- **Screening.** Screening is the process of evaluating and selecting human and media sources for the prioritized collection of information in support of commander's PIRs and IRs. Screening categorizes and prioritizes sources based on the probability of a particular source having priority information and the degree of difficulty of extracting that information from the source. While screening is not in itself an information collection technique, it is vital to the rapid collection of information.
- **Liaison.** HUMINT elements conduct liaison with US, multinational, and HN military and civilian agencies, to include NGOs. Liaison is conducted to obtain information of interest and to coordinate or deconflict HUMINT activities.
- **HUMINT Contact Operations.** HUMINT contact operations are tactically oriented, overt collection activities that use human sources to identify attitude, intentions, composition, strength, dispositions, tactics, equipment, target development, personnel, and capabilities of those elements that pose a potential or actual threat to US and multinational forces. These forces provide early warning of imminent danger to US and multinational forces and contribute to the MDMP.
- **Document Exploitation (DOCEX).** DOCEX is the systematic extraction of information from all media formats in response to collection requirements.
- **Interrogation.** Interrogation is the systematic effort to procure information to answer specific collection requirements by direct and indirect questioning techniques of persons who are in the custody of the forces conducting the questioning.

UNIT SUPPORT TO HUMINT COLLECTION

6-5. Small units contribute to HUMINT collection through a number of different ways.

Tactical Questioning

6-6. Tactical questioning can provide critical information for situational understanding. Tactical questioning is the expedient initial questioning for information of immediate tactical value. Soldiers conduct tactical questioning based on the unit's SOP, ROE, and the order for that mission. Unit leaders must include specific guidance for tactical questioning in the order for appropriate missions. The unit S3 and S2 must also provide specific guidance down to the unit level to help guide tactical questioning. For a more detailed discussion of tactical questioning, see ST 2-91.6.

Combat Patrolling

6-7. Through observing and interacting with the local environment during the conduct of missions, handling EPWs/detainees, and handling captured documents, soldiers serve as the commander's "eyes and ears" whether—
- Performing traditional offensive or defensive operations.
- Performing a patrol in a stability operation.
- Manning a TCP or a roadblock in a support operation.
- Occupying an observation post.
- Passing through an area in a convoy.
- Performing any operation that involves observing and reporting on elements of the environment and activities in the AO.

ISR Operations

6-8. The information that the soldier reports as a result of completed missions up the chain of command forms a vital part of planning and operations. Careful and quick handling of EPWs/detainees and documents also helps the ISR effort. Unit headquarters must ensure that the reports they receive and forward contain all the subtle detail provided by the small unit. For tactical operations, there are four levels of reporting:
- Immediate reporting of information of critical tactical value, based either on predetermined criteria or common sense.
- Normal reporting, submitted before the unit S2 section performs the debriefing.
- Debriefing performed by the S2 section.
- Follow-up reporting, submitted after the unit S2 section performs the debriefing.

6-9. The four levels of reporting facilitate the unit S2 section capturing all the fine details of the small unit's activities for all-source analysis and future planning. The unit S2 must ensure that information of HUMINT and CI value is reported to the G2X.

ANALYSIS

6-10. IPB, all-source, and single-source analysis are used to template adversarial OB. HUMINT elements analyze operational taskings to determine the best collection methods to employ in satisfying CCIRs (PIRs and FFIRs) and IRs. As information is collected, it is analyzed for completeness and accuracy to identify significant facts for subsequent interpretation and inclusion in intelligence products, and passing to other collectors. This analysis also helps to identify collection gaps and focus or refocus collection efforts. Analysis provides the commander with situational understanding of the battlespace to execute operations. Raw information, open-source material, and finished intelligence products are analyzed in response to local and national requirements. Analysis occurs at all levels from tactical to strategic.
- At the tactical level, HUMINT teams focus their efforts on supporting mission requirements and contributing to the all-source COP.

- Operational analysis is used to assess adversary intentions, capabilities, dispositions, and their regional impacts.
- Strategic analysis supports national and Army programs through compilation of local and regional analysis into global assessments; identifies technology development trends and patterns in military activities and capabilities.

6-11. HUMINT products consist of, but are not limited to, target nomination, input to threat and vulnerability assessments, intelligence estimates, and intelligence information reports. Finalized intelligence derived from HUMINT activities is incorporated into joint and national intelligence databases, assessments, and analysis products. HUMINT products are also incorporated into the COP to support situational awareness. HUMINT production takes place at all levels.

- Operational and tactical production includes tactical alerts, spot reports, and current intelligence; input to threat and/or vulnerability assessments tailored to specific activities, units, installations, programs or geographic areas, and target studies to support contingency planning and major exercises; studies of military activities and capabilities.
- Strategic products include assessments supporting national and Army information requirements on foreign technology development; worldwide assessments of the organization, location, funding, training, operations capabilities and intentions of terrorist organizations; analyses of the capabilities of international narcotics trafficking organizations.

OPERATIONAL EMPLOYMENT

6-12. Army HUMINT must support the full spectrum of military operations. Commanders require a well-trained HUMINT force consisting of AC, RC, civilian government, and contractor personnel. The HUMINT force is focused on and dedicated to the collection of data and information relevant to the commander's PIRs and IRs. Effective employment of Army HUMINT elements in all phases of operations and at all levels from tactical to strategic will remain paramount to ensuring that commanders have the best pictures possible of their adversaries. HUMINT Collectors use the following collection methods.

LEVELS OF EMPLOYMENT

6-13. **Strategic and Departmental.** Strategic and departmental operations will be conducted by HUMINT elements supporting national, DOD, and DA required missions (for example, support to NATO and special operations and missions). The Army strategic HUMINT force resides in DHS. Trained Army HUMINT professionals must be available to augment DOD HUMINT efforts and coordinate between Army and Joint or DOD elements during crisis and war. Strategic HUMINT activities are generally carried out by departmental level HUMINT elements or Army HUMINT assets directly assigned to DOD and Joint positions. The Army overt HUMINT capability at the strategic level will be deployed and under operational control (OPCON) of the Army Service Component Commander (ASCC).

6-14. **Operational.** HUMINT operational missions must support combatant commanders, generally in geographic AORs. Operational HUMINT assets provide capabilities to support theater requirements for HUMINT collection. Operational HUMINT elements will focus on threat identification and capabilities. As with all levels of employment, HUMINT activities and functions will include DOCEX, collection, operations, analysis, and production. HUMINT elements must be capable of quickly transitioning from a peacetime mission to crisis operations to support combatant commander requirements. Operational elements may also be deployed to support or reinforce tactical forces in CONOPS.

6-15. **Tactical.** Army HUMINT play a crucial role in supporting tactical forces. HUMINT teams conduct operations throughout the battlespace during CONOPS. CONOPS support activities include debriefings, screenings, contact operations, and interrogations. HUMINT activities in CONOPS focus on the threat and assisting the senior intelligence officer (SIO) and commander in understanding the threat's decisionmaking process. During peacetime, organic tactical HUMINT teams conduct activities pursuant with approved regulations and command guidance.

SUPPORT TO CONTINGENCY OPERATIONS

6-16. **Initial Phase.** The initial phase of operations from peacetime military engagement (PME) to major theater war (MTW) lays the foundation of future team operations. In general, the priority of effort focuses inward on security of operating bases, areas of troop concentration, and C2 nodes to identify the collection threat to US forces that could be used by adversarial elements to plan hostile acts against US activities and locations.

6-17. **Continuation Phase.** Once security of the operating bases has been established, the operational focus of HUMINT teams shifts outside the operating base to continue to detect, identify, and neutralize the collection threat to US forces as well as to provide I&W of hostile acts targeting US activities. The HUMINT team uses several collection methods, to include HUMINT contact operations, elicitation, and liaison, to answer the supported commander's requirements.

6-18. **HUMINT Team.** A key element to the HUMINT team's success is the opportunity to spot, assess, and develop relationships with potential sources of information. Operating as independent teams, without being tied to combat assets, enables the HUMINT team's maximum interaction with the local population, thereby maximizing the pool of potential sources of information. A second key element of a HUMINT team's success is its approachability to the local population. A soft posture enables a HUMINT team to appear as non-threatening as possible. Experience has shown that the local population in general is apprehensive of fully and openly armed patrols and soldiers moving around population centers. During some operations civilian attire or non-tactical vehicles may be used to lower the

HUMINT team profile. NOTE: In some special situations, these measures are taken to make the operation less visible to the casual observer. Also, in some cultures, sharing food and beverages among friends is expected; exceptions to restrictions or general orders should be considered to facilitate successful HUMINT team operations, many of which are geared towards developing relationships with potential sources of information.

SUPPORT TO INSTALLATIONS AND OPERATING BASES

6-19. Commanders may restrict personnel to base camps and installations during the initial stages of operations (PME to MTW), when the operational environment is being assessed, or as a temporary expedient when the threat level exceeds the ability to provide reasonable FP. Operational restrictions minimize the risk to the HUMINT team, but minimizing its collection potential may increase the risk to the force as a whole. While confined to an installation or a base camp, the HUMINT team can maintain a limited level of information collection by—

- Screening locally employed personnel.
- Debriefing combat and ISR patrols.
- Debriefing friendly force personnel who are in contact with the local population.
- Conducting limited local open-source information collection.
- Contributing to the threat and vulnerability assessments of the base camp.

TACTICS, TECHNIQUES, AND PROCEDURES

6-20. At the HUMINT team level, team members conduct mission analysis and planning specific to their AO. Backwards planning and source profiling are used extensively to choose HUMINT targets. To verify adequate area coverage, the HUMINT team may periodically develop and use HUMINT target overlays and other HUMINT analytical tools that illustrate the HUMINT situation, identify gaps, and help refocus the collection effort.

6-21. The HUMINT team is also in constant contact with the supported S2 and the other ISR assets (Scouts, PSYOP, CA, and MPs) in order to coordinate and deconflict operations and to cross-check collected information. The supported unit S2, with the help of the HUMINT team, regularly and systematically debriefs all ISR assets.

6-22. The HUMINT team must be integrated into the supported unit's ISR plan. The HUMINT operational management team (OMT) chief will advise the supported unit on the specific capabilities and requirements of the team to maximize mission success.

OPERATIONAL RISK MITIGATION

6-23. The employment of HUMINT teams includes varying degrees of contact with the local population. As the degree of contact with the population increases, both the quantity and quality of collection increases. In many instances, however, there is a risk to the team inherent with increased exposure to the local population. The decision at what level to employ a team is METT-TC dependent. The risk to HUMINT assets must be balanced with the need to collect priority information and to protect the force as a whole. ROE, SOFA, direction from higher headquarters, and the overall threat level may also restrict the deployment and use of HUMINT teams. The commander should consider exceptions to the ROE to facilitate HUMINT collection.

6-24. Risks are minimized through the situational awareness of HUMINT team members. They plan and rehearse to readily react to any situation and carry the necessary firepower to disengage from difficult situations. If it becomes necessary to call for assistance, adequate and redundant communications equipment is critical. These scenarios and actions should be trained prior to deployment into a contingency area and rehearsed continuously throughout the deployment.

6-25. A supported unit commander is often tempted to keep the HUMINT team "inside the wire" when the threat condition (THREATCON) level increases. The supported commander must weigh the risk versus potential information gain when establishing operational parameters of supporting HUMINT teams. This is necessary especially during high THREATCON levels when the supported unit commander needs as complete a picture as possible of the threat arrayed against US and multinational forces.

6-26. When it is not expedient to deploy the HUMINT team independently due to threat levels or other restrictions, the team can be integrated into other ongoing operations. The HUMINT team may be employed as part of a combat, ISR, or MP patrol or used to support CA, PSYOP, engineer, or other operations. This method reduces the risk to the team while allowing a limited ability to collect information. It has the advantage of placing the team in contact with the local population and allowing it to spot, assess, and interact with potential sources of information. However, this deployment method restricts collection by subordinating the team's efforts to the requirements, locations, and timetables of the unit or operation into which it is integrated and does not allow for the conduct of sensitive source operations. **This method of employment should be considered a last resort.**

HUMINT EQUIPMENT

6-27. Basic C2, transportation, and weapons requirements do not differ significantly from most soldier requirements and are available as unit issue items. However, HUMINT teams have unique communications, collection, processing, and mission-specific requirements.

COMMUNICATIONS

6-28. Dedicated and secure long-range communications are key to the success of the HUMINT team mission. HUMINT team operations require a secure, three-tiered communications architecture consisting of inter/intra-team radios, vehicle-based communications, and a CI and HUMINT base station.

6-29. The HUMINT team must have access to existing communications networks such as the tactical local area network (LAN). The HUMINT team must also be equipped with its own COMSEC devices. It is imperative that the HUMINT team acquire access to the public communication system of the HN. This can be in the form of either landlines or cellular telephones. Such access enables the HUMINT team to develop leads which can provide early indicators to US forces.

Interoperability

6-30. Communications systems must be equipped with an open-ended architecture to allow for expansion and compatibility with other service elements, government organizations, NGOs, and multinational elements to effectively communicate during CONOPS. All ISR systems must be vertically and horizontally integrated to be compatible across all BOSs and with Legacy and Interim Force elements.

Satellite Communications On-The-Move (SOTM)

6-31. To provide real time or NRT information reporting, HUMINT elements must have the capability to transmit voice, data, imagery, and video while on the move. HUMINT teams must be able to transmit while geographically separated from their parent unit while operating remotely. This broadband requirement can only be achieved through a SATCOM capability and must be achievable while mobile.

HUMINT COLLECTION AND PROCESSING SYSTEMS

6-32. The HUMINT team must rely on automation to achieve and maintain information dominance in a given operation. With time, effective collection planning and management at all echelons, the HUMINT team can collect a wealth of information. The sorting and analysis of this information in a timely and efficient manner is crucial to operations. Automation helps the HUMINT team to report, database, analyze, and evaluate the collected information quickly and to provide the supported unit with accurate data in the form of timely, relevant, accurate, and predictive intelligence.

6-33. Automation hardware and software must be user friendly as well as interoperable among different echelons and services. They must interface with the communications equipment of the HUMINT team as well as facilitate the interface of audiovisual devices. Technical support for hardware and software must be available and responsive.

6-34. The demand for accurate and timely HUMINT reporting, DOCEX, and open-source information has grown tremendously. Biometric (physiological, neurological, thermal analysis, facial and fingerprint recognition) technologies will allow rapid identification; coding and tracking of adversaries and human sources; and cataloging of information concerning enemy prisoners of war (EPWs), detainees, and civilians of HUMINT interest on the battlefield. Biometrics will also provide secure authentication of individuals seeking network or facility access.

6-35. HUMINT teams work with multinational forces, and other foreign nationals, and require the ability to communicate in their respective languages. Often HUMINT personnel have little or no training in the target language, and lack of skilled interpreters can hinder HUMINT activities. HUMINT teams require textual and voice translation devices, source verification, and deception detection machines (biometrics) to improve collection capability and accuracy.

6-36. HUMINT teams require dynamic machine language translation (MLT) tools that provide both non-linguists and those with limited linguist skills a comprehensive, accurate means to conduct initial screenings and basic interviews in a variety of situations. HUMINT elements will focus on in-depth interviews and communications with persons of higher priority. MLT tools will minimize reliance on contract linguists and will allow soldiers to concentrate on mission accomplishment.

MISSION SPECIFIC

6-37. The HUMINT team may conduct night operations and must be equipped with NVDs for its members and photographic and weapons systems. The HUMINT team also may operate in urban and rural areas, where the threat level can vary from semi-hostile to hostile. The safety of the HUMINT team can be enhanced with equipment that can detect, locate, suppress, illuminate, and designate hostile optical and E-O devices. In addition, high power, gyro-stabilized binoculars, which can be used from a moving vehicle, also increase the survivability of the HUMINT team. It also gives the team another surveillance and collection device.

6-38. Some of the HUMINT team missions may require the documentation of incidents. The teams can use the following equipment in their open-source collection efforts.

- Small, rugged, battery-operated digital camcorders and cameras which are able to interface with the collection and processing systems as well as communication devices.
- GPSs that can be mounted and dismounted to move in the AO efficiently.
- Short-range multichannel RF scanning devices that can also identify frequencies which enhance their security.

6-39. In some cases HUMINT teams require a stand-off, high resolution optical surveillance and recording capability that can provide target identification at extended ranges to protect the intelligence collector while avoiding detection by the adversary target. An advanced optical capability provides intelligence collectors the ability to locate and track adversarial targets (passive and hostile) for identification, collection, and target exploitations.

INTEGRATION OF LINGUISTS

6-40. Integrating linguists into the HUMINT team should take place as soon as possible. Security clearances and contractual agreements will help the team determine the level of integration.

6-41. Along with the basic briefing of what is expected of the civilian linguists as interpreters, HUMINT teams should be informed about the civilians' chain of command and the scope of their duties beyond interpreting. The HUMINT team leader must ensure that linguists are trained and capable of completing all tasks expected of them.

FM 2-0

BATTLE HAND-OFF

6-42. HUMINT teams are always engaged. A good battle hand-off is critical to smooth transition and mission success. The battle hand-off can directly contribute to mission success or failure of the outgoing team, but especially of the incoming team. The battle hand-off begins the first day the HUMINT team begins to operate in an AO. Regardless of how long the team believes it will operate within the AO, it must ensure there is a seamless transition to an incoming team, other US unit, or agency. The HUMINT team accomplishes this transition by establishing procedures for source administration, database maintenance, and report files.

6-43. Teams must plan and implement a logical and systematic sequence of tasks to enable an incoming team to assume the operations in the AO. Adequate time must be allotted for an effective battle hand-off. In some environments, a few weeks may be necessary to accomplish an effective battle hand-off. Introductions to sources of information, especially HUMINT contact operations sources, are critical and teams must prioritize their time. During this time the outgoing HUMINT team must familiarize the new HUMINT team with all aspects of the operation, which includes past, present, and planned activities within the AO. Area orientation is critical. These include major routes, population centers, potential hot spots, and other points of interest (such as police stations, political centers, and social centers).

ORGANIZATION

6-44. HUMINT activities require a complex C2 relationship to ensure that the requirements of the supported commander are fulfilled while balancing the need for strict integrity and legality of HUMINT operations. This complex relationship balances the role of the SIO as the requirements manager and the *2X as the mission manager with the MI commander as the asset manager. ("*2X" indicates 2X functions at all levels.)

COMMAND VERSUS CONTROL

6-45. ARFOR will normally deploy as part of a joint and/or multinational operation. In all cases, commanders at each echelon will exercise command over the forces assigned to their organization. Command includes the authority and responsibility for effectively using resources, planning for and employment of forces, and ensuring that forces accomplish assigned missions. Leaders and staffs exercise control to facilitate mission accomplishment.

6-46. While the MI commander supervises subordinates and produces reports, the *2X synchronizes activities between intelligence units and provides single-source processing and limited analysis. While the MI commander takes care of the operators executing missions, the *2X obtains the data and reports from higher echelons required to execute the missions.

STAFF RESPONSIBILITIES AND FUNCTIONS

*2X Staff

6-47. The *2X staff is responsible for the integration, correlation, and fusion of all Human Sensor information into the Intelligence BOS within the *2X

AOIR. The *2X is also responsible for analyzing adversary collection, terrorist and sabotage activities, developing countermeasures to defeat threat collection activities, identifying and submitting collection requirements to fill collection gaps, and providing input to the all-source picture regarding adversary intelligence activities.

*2X Staff Officer

6-48. The *2X Staff Officer provides CI and HUMINT collection expertise. The *2X—

- Is the single focal point for all matters associated with CI and HUMINT in the AOIR.
- Is the CI and HUMINT advisor to the G2 and commander.
- Is an extension of the collection manager and ensures that the best asset or combinations of assets are used to satisfy information requirements.
- Along with his subordinate elements (CICA, HOC, OSC, CI Analysis Cell [CIAC], and HAC), exercises technical control over his assigned Army CI and HUMINT elements in the designated AOIR.
- Is the principal representative of the G2 and the commander when coordinating and deconflicting CI and HUMINT activities with national or theater agencies operating in the AOIR.
- Supports specific RM efforts in conjunction with the requirements manager through—
  - Planning and coordinating CI and HUMINT operations.
  - Reviewing and validating HUMINT requirements.
  - Recommending assignment of tasks to specific collectors.
  - Conducting liaison with non-organic HUMINT collection. This liaison includes national level and multinational force assets for source deconfliction and special activities outside the *2X AOIR.
- Will provide OMTs with capability to reach back to current database information, technical information and guidance, and source deconfliction necessary to monitor the collection activities of the HUMINT teams.

The HUMINT Analysis Cell (HAC)

6-49. The HAC is the single-source fusion point for all HUMINT reporting and operational analysis. It determines gaps in reporting and coordinates with other analysis teams and technical controllers to cross-cue other collection sensor systems. The HAC—

- Uses analytical tools to develop long-term collection plans and provides reporting feedback that supports all HUMINT and CI entities in the supported command's AOIR.
- Produces and disseminates HUMINT products and provides input to intelligence summaries.
- Uses analytical tools found at the ACE or JISE to develop long-term analyses and provides reporting feedback that supports the J/G/S2X, HUMINT operations section, OMTs, and HUMINT teams.
- Produces country and regional studies tailored to HUMINT collection.

- Compiles target folders to assist J/G/S2X assets in focusing collection efforts.
- Analyzes and reports on trends and patterns found in HUMINT reporting.
- Analyzes source reliability and credibility as reflected in reporting and communicating that analysis to the collector.
- Develops and maintains databases specific to HUMINT collection activities that directly support the collection efforts of HUMINT teams and are directly accessible by HUMINT teams.
- Provides collection requirements input to the HOC.
- Supports RM through the development of HUMINT SIRs based on command PIRs.
- Answers HUMINT-related RFIs.

6-50. For intelligence reach operations, HUMINT products are available and disseminated in a variety of forms. It is incumbent on the requestor to ensure that the HUMINT product can be transmitted over the available communications systems. This includes verifying the appropriate security level of the communications systems.

HUMINT Operations Cell (HOC)

6-51. The HOC in the *2X coordinates and synchronizes all HUMINT activities in the AOIR. The HOC exercises technical control over all HUMINT entities in the designated AOIR and deconflicts HUMINT activities with higher, lower, and adjacent HUMINT elements. The HOC accomplishes all responsibilities through coordination with the operational units and the CICA and operations support cell (OSC). The HOC tracks all HUMINT activities in the AOR. The J/G2X uses this information to advise the SIO on all HUMINT activities conducted within the AO. The HOC—

- Exercises technical control of all HUMINT assets and coordinates and deconflicts HUMINT activities in the deployed AO.
- Establishes and maintains a HUMINT source database.
- Coordinates and supervises HUMINT FP source operations conducted by all services and components in the AO.
- Develops and manages collection requirements for HUMINT in coordination with the requirements manager.
- Develops and provides the HUMINT portion of the intelligence synchronization plan to the J/G/S2X and requirements manager for inclusion in the intelligence synchronization plan.
- Coordinates the activities of HUMINT collectors assigned or attached to interrogations and debriefing facilities.
- Expedites preparation of intelligence information reports and their distribution to consumers at all levels.
- Performs liaison with HN and US national HUMINT organizations.

Operations Support Cell

6-52. The OSC in the *2X staff maintains the source registry for all HUMINT activities in the designated AOIR. The OSC provides management of intelligence property book operations, source incentive programs, and ICFs.

6-53. Integrating linguists and DOD emergency-essential civilians, such as technical support contractors, into the HUMINT team should take place as soon as possible. Security clearances and contractual agreements will help the team determine the level of integration.

6-54. Along with the basic briefing of what is expected of the civilian linguists as interpreters and the emergency-essential civilians as support personnel, HUMINT teams should be informed about the civilians' chain of command and the scope of their duties beyond interpreting and technical support. The HUMINT team leader must ensure that linguists and emergency-essential civilians are trained and capable of completing all tasks expected of them.

Counterintelligence Coordinating Authority

6-55. The CICA coordinates all CI activities for a deployed force. There can be only one CICA in a theater of operations. When multiple echelons exist, the highest echelon has the CICA and subordinate G2X offices have a CICA. Depending on the size and scope of the operation, the CICA could be the unified command's CI staff officer; the CI or HUMINT staff officer from corps or division; or a senior warrant officer or branch CI officer designated by the unified command CI staff officer or task force commander. For more information about the CICA functions, see Chapter 11.

HUMINT TEAM STRUCTURE

Operational Management Team

6-56. The OMT is a four-person team consisting of a warrant officer (WO), two noncommissioned officers (NCOs), and a junior enlisted soldier. Civilians may be inserted into this structure when appropriate. Rank structure and standards of grade for OMTs vary depending upon the skill sets required and mission focus. HUMINT OMTs provide operational guidance for two to four HUMINT teams, depending on mission focus and operational tempo. When two or more HUMINT teams are deployed in DS of a maneuver element, an OMT also deploys to provide technical control. The OMT works closely with the supported S2 and analysis and control team (ACT) to furnish current threat information and answer the supported commander's PIRs and IRs. OMTs coordinate with the supported 2X and manage subordinate HUMINT team echelons to—

- Provide guidance and technical control of operational activity.
- Provide the collection and operational focus for HUMINT teams.
- Provide quality control and dissemination of reports for subordinate HUMINT teams.
- Conduct single-discipline HUMINT analysis, and assist in mission analysis for the supported commander.

- Act as a conduit between subordinate HUMINT teams, the HOC, and supported unit headquarters.
- Provide administrative support for subordinate HUMINT teams, to include reporting mission and equipment status to the HOC and the supported unit headquarters.
- Educate the supported commander on the capabilities of the HUMINT teams.
- Integrate the HUMINT teams directly into the maneuver commander's ISR planning.

HUMINT Team

6-57. The HUMINT team is a four-person team consisting of two NCOs and two junior enlisted personnel. Civilians may be inserted into this structure when appropriate. Rank structure and standards of grade for HUMINT teams vary depending upon the skill sets required and the mission focus. HUMINT teams are trained to execute the full range of HUMINT functions as defined in **HUMINT FUNCTIONS** section. However, they may be assigned to mission-focused elements such as DOCEX, interrogation, debriefing, or contact operations.

Chapter 7

Imagery Intelligence

DEFINITION

7-1. IMINT is intelligence derived from the exploitation of imagery collected by visual photography, infrared, lasers, multi-spectral sensors, and radar. These sensors produce images of objects optically, electronically, or digitally on film, electronic display devices, or other media.

ROLE

7-2. The role of imagery is to assist the commander in focusing and protecting his combat power. Imagery often enhances the commander's situational understanding of the battlespace. Other than direct human observation, imagery is the only intelligence discipline that allows the commander to see the battlefield in real time as the operation progresses. In those cases where maps are not available, digital imagery in hardcopy or softcopy can be used as a substitute. Imagery can also be used to update maps or produce grid-referenced graphics. Detailed mission planning often requires imagery, to include three-dimensional stereo images, in order to provide the degree of resolution necessary to support such specialized planning.

FUNDAMENTALS

7-3. Some imagery assets are very responsive to the individual commander's intelligence requirements. Some imagery systems can directly transmit imagery into the tactical operations center (TOC); examples include imagery from UAVs and the Joint Surveillance Target Attack Radar System (JSTARS). This direct downlink enables the G2/S2 to use the imagery as soon as possible instead of having to wait for finished imagery products. A note of caution is required, however, if not a trained imagery analyst because the G2/S2 could incorrectly interpret the imagery.

7-4. Imagery-related equipment has undergone a reduction in size as well as a reduction in the time it takes to provide products, particularly softcopy imagery. The modularity and size reduction of imagery analysis, processing, and display systems make transport easier; they also allow the commander to bring lesser amounts than in the past, while still retaining those systems (or subsystems) required to complete the mission. Additionally, data compression allows faster transmission of imagery products directly to the warfighter.

SOURCES OF IMAGERY

7-5. There are three general sources of imagery: national, civil, and commercial. National imagery traditionally refers to imagery collected by DOD imagery systems. However, there are other sources of imagery provided

FM 2-0

by non-national sources such as handheld photography (film, digital, and video), UAV imagery, and gun-camera images.

National

7-6. National systems are developed specifically for supporting the President of the United States, the Secretary of Defense, other national agencies, and US military forces. These systems respond to the needs of the nation and those of the combatant commands.

Civil

7-7. Civil imagery systems are usually government funded in terms of building, launching, and operating the system. In many, but not all, cases the agencies operating these civil imagery systems also process, distribute, and archive the imagery data or images.

Commercial

7-8. Commercial companies build, launch, and operate imagery systems for profit. In times of crises, license agreements with the US Government obligate US commercial satellite imaging systems to provide data only to the US Government at the market value. This protects information concerning US operations from adversarial exploitation from commercial systems such as the ACE Imagery Company. However, foreign commercial imagery systems are not bound to this arrangement, and thus may be used by our nation's adversaries. Commercial imagery has become increasingly valuable for many reasons:

- Due to its unclassified nature, civil and commercial imagery is useful in an open environment, may be released to multinational partners, and can be made available to the press. They are especially useful for geospatial products. Their use allows national systems more time to focus on other intelligence functions.
- Civil and commercial imagery offer radar and multi-spectral. Some offer large area collection useful for broad area coverage purposes.
- Commercial satellite imagery resolution varies from less than one meter to several kilometers.

7-9. The Central Imagery Tasking Office (CITO) is responsible for ordering commercial imagery. The Commercial Satellite Imagery Library is available to research DOD purchased commercial imagery. The G2/S2 should consult the CITO when forming commercial imagery requests. NGA will deliver the imagery primarily on CD-ROM media via courier or mail service. Limited digital or electronic delivery is available as well.

TYPES OF IMAGERY SENSORS

7-10. There are four types of imagery sensors: visible (optical), infrared, radar, and multi-spectral. Each sensor has a unique capability, with distinct advantages and disadvantages. The G2/S2 must understand each sensor's capability in order to select the best sensor for the mission and thus enable the user to better understand the intelligence received. Certain sensors are better suited for military operations than others. (See Table 7-1.)

Table 7-1. Sensor Characteristics Matrix.

SENSORS	ADVANTAGES	DISADVANTAGES
Visible (Optical) Best tool for daytime, clear weather, detailed analysis. Includes video and electro-optical.	• Affords a familiar view of a scene. • Offers system resolution that cannot be achieved in other optical systems or in thermal images and radars. • Preferred for detailed analysis and mensuration. • Offers stereoscopic viewing.	• Restricted by terrain and vegetation. • Limited to daytime use only. • Reduced picture size.
Infrared Best tool for nighttime, clear weather, detailed analysis. Includes Overhead Non-Imaging Infrared (ONIR).	• A passive sensor and is impossible to jam. • Offers camouflage penetration. • Provides good resolution. • Nighttime imaging capability.	• Not effective during thermal crossover periods. • Product not easily interpretable. • Requires skilled analysis. • Cannot penetrate clouds.
Radar Useful for detecting presence of objects at night and in bad weather. Includes synthetic aperture radar (SAR), coherent change detection (CCD), and MTI.	• All weather; can penetrate fog, haze, clouds, smoke. • Day or night use. • Does not rely on visible light nor thermal radiation. • Good standoff capability. • Large area coverage. • Allows moving target detection. • Foliage and ground penetration.	• Product not easily interpretable. • Requires skilled analysis. • Terrain masking inhibits use.
Multi-Spectral Imagery (MSI) Best tool for mapping purposes and terrain analysis.	• Large database available. • Band combinations can be manipulated to display desired requirements. • Images can be merged with other digital data to provide higher resolution.	• Product not easily interpretable. • Requires skilled analysis. • Computer manipulation requires large amounts of memory and storage; requires large processing capabilities.

PLAN

7-11. The first step in planning for IMINT is determining the need for IMINT products based on the PIRs and the initial IPB. The G2/S2 should research targets using online imagery databases early and request those imagery products that are not perishable for contingency planning. National and COCOM imagery databases may hold recently imaged targets that could meet the commander's immediate needs instead of requesting new imagery. The staff must clearly articulate their intelligence requirements to include communicating what the mission is and how the requested product will aid in mission accomplishment. The G2/S2 should submit the imagery requirement using established procedures such as those in the unit's SOP or as established by the COCOM.

7-12. The G2/S2 must also determine the specific imagery requirements so as not to burden the system with unnecessary requests. The desire for imagery products often exceeds the capabilities of the IMINT system. Therefore, it is imperative that the G2/S2 consider what type of analysis they require, and request only that which they require. The specifications of the request for IMINT products often affect the timeliness of the response. For example, determining if vehicles are tanks takes less time and requires less resolution than determining if a tank is a T-64 or a T-72.

7-13. Here are some of the roles IMINT can perform that the G2/S2 may consider when determining IMINT requirements:

- Imagery can detect and/or identify and locate specific unit types, equipment, obstacles, potential field fortifications, etc., from which intelligence analysts are able to analyze enemy capabilities and develop possible COAs.
- Imagery can also update maps and enhance the interpretation of information from maps. Detailed mission planning uses imagery to include stereo images for three-dimensional viewing of the terrain and many other geospatial uses.
- Imagery can be used as a substitute for maps when maps are not available. The most common application of this technique is by constructing imagery mosaics: a combination of two or more overlapping photographic prints that form a single picture.
- MTI displays or products can provide an NRT picture of an entity's movement by indicating its speed, location, and direction of travel. MTI systems do not differentiate friendly from enemy. Imagery assets, particularly MTI systems, are useful in cueing other ISR systems.
- Imagery can be used to support protection of the force by helping the commander visualize how his forces look—including their disposition, composition, and vulnerabilities—as exploited by enemy IMINT systems.
- Imagery analysts use combat assessment imagery to confirm destruction, determine the percentage of destruction, or whether the target was unaffected.

PREPARE

7-14. The G2/S2 IMINT-related actions during the prepare function of the intelligence process include establishing or verifying the portion of the intelligence communications architecture that supports IMINT display and analysis functions properly. Additionally, the G2/S2 must ensure that required IMINT analytical assets and resources are prepared to provide support or are available through intelligence reach. Lastly, the G2/S2 must also ensure IMINT reporting and dissemination channels and procedures are in place and rehearsals are conducted with all pertinent IMINT elements to ensure interoperability.

FM 2-0

COLLECT

7-15. As previously mentioned, there are four types of imagery sensors. Depending on the type of sensor, it can record hardcopy or softcopy single frame or continuous (video). A given imagery target will not necessarily receive continuous coverage due to the possible conflict between the number and priority of targets and the number and availability of imagery assets. However, a commander may decide to have continuous surveillance of certain targets, for specified periods of time, usually using his own imagery assets (for example, UAV) even though this detracts from the commander's ability to use these assets for other imagery targets within his AOI.

PROCESS

7-16. The process function regarding IMINT involves converting imagery data into a form that is suitable for performing analysis and producing intelligence. Examples of IMINT processing include developing film, enhancing imagery, converting electronic data into visual displays or graphics, and constructing electronic images from IMINT data.

PRODUCE

7-17. The IMINT producer must ensure the IMINT product satisfies the associated intelligence requirements and that the product is in the required format. The quality and resolution of the product is highly dependent upon the type of sensor, the time of day, and the weather conditions, as well as the imagery analyst's ability to identify items, vehicles, equipment, and personnel within the images. Specific IMINT products are discussed in the **ANALYZE** section below.

ANALYZE

7-18. Timeliness is critical not only to IMINT collection but also to IMINT analysis and reporting. It is difficult to separate IMINT reporting from IMINT analysis in this discussion. This is demonstrated by the three phases of IMINT reporting presented below; all are dependent upon the timeliness requirements. Each phase represents a different degree of analysis and period of time available to accomplish the exploitation of the imagery.

- First Phase imagery analysis is the rapid exploitation of newly acquired imagery and reporting of imagery-derived information within a specified time from receipt of imagery. This phase satisfies priority requirements of immediate need and/or identifies changes or activity of immediate significance. First Phase imagery analysis results in an Initial Phase Imagery Report (IPIR).
- Second Phase imagery analysis is the detailed exploitation of newly acquired imagery and the reporting of imagery-derived intelligence and information while meeting the production and timeliness requirements. Other intelligence discipline source material may support phase two imagery as appropriate. Second phase imagery analysis results in a Supplemental Imagery Report (SUPR).
- Third Phase imagery analysis is the detailed analysis of all available imagery pertinent to an SIR, and the subsequent production and reporting resulting from this analysis within a specified time. This

phase provides an organized detailed analysis of an imagery target or topic, using imagery as the primary data source but incorporating data from other sources as appropriate.

7-19. The two types of imagery exploitation are national and direct support (DS).

- National exploitation is imagery exploitation that supports presidential requirements, National Security Council (NSC) requirements, congressional requirements, or requirements of a common concern to the intelligence community.
- DS exploitation is imagery exploitation that supports assigned missions of a single agency, department, or command (the warfighter).

7-20. IMINT assets will complete DS exploitation in order to satisfy (First Phase) requirements and report the results as soon as possible, but not later than 24 hours after receipt of the imagery. Collectors will complete national exploitation in order to satisfy (Second and Third Phases) requirements and report the results within the time specifications of each individual requirement.

DISSEMINATE

7-21. IMINT products are distributed or disseminated in hardcopy, softcopy, or direct viewing such as through remote terminals. The distribution of hardcopy products will be via couriers or other type of mail system. The dissemination of softcopy products will be either as hardcopy products (for example, CD-ROM and 3.5-inch disks) or electronically. The requestor must ensure that the requested product is transmittable over the available communications systems.

ASSESS

7-22. The requestor should immediately assess the imagery product upon receipt for accuracy and relevance to the original request. The requestor must then notify the producer and inform him of the extent to which the product answered the PIR. Providing feedback to the producer regarding the product helps ensure the producer will provide the required information in the correct format. The following are some of the questions which the requestor should consider when providing feedback to the producer.

- Is the format of the product acceptable?
- Is additional information needed on the product or future products?
- Is excess information included on the product?

Chapter 8
Signals Intelligence

DEFINITION

8-1. SIGINT is a category of intelligence comprising either individually or in combination all COMINT, ELINT, and FISINT, however transmitted; intelligence is derived from communications, electronics, and foreign instrumentation signals. SIGINT has three subcategories:
- COMINT – The intelligence derived from foreign communications by other than the intended recipients.
- ELINT – The technical and geo-location intelligence derived from foreign non-communications electromagnetic radiations emanating from other than nuclear detonations or radioactive sources.
- FISINT – Technical information and intelligence derived from the intercept of foreign electromagnetic emissions associated with the testing and operational deployment of non-US aerospace, surface, and subsurface systems. Foreign instrumentation signals include but are not limited to telemetry, beaconry, electronic interrogators, and video data links. (See JP 1-02)

8-2. SIGINT provides intelligence to the commander based upon intercepted communications and provides transmitter location data.

ROLE

8-3. SIGINT provides intelligence on threat capabilities, disposition, composition, and intentions. In addition, SIGINT provides targeting information for the delivery of lethal and non-lethal effects.

FUNDAMENTALS

8-4. It is important that the G2/S2 understand how SIGINT assets are organized not only within the Army but also throughout the DOD. The majority of SIGINT assets are located at EAC. SIGINT assets from all the armed services, combined with national SIGINT assets, work together to support commanders from the tactical to the strategic level. Only by understanding the SIGINT structure that transcends traditional service component boundaries can the G2/S2 understand how to use SIGINT effectively.

TECHNICAL CONTROL AND ANALYSIS ELEMENT (TCAE) ARCHITECTURE

8-5. The TCAE architecture supports Army SIGINT collectors and analysts by providing SIGINT-specific intelligence and guidance to SIGINT personnel. The SIGINT technical architecture complements existing C2 relationships; it does not replace the commander's authority or chain of command.

8-6. There is a TCAE at each Army operational echelon. The TCAE at each echelon provides a single POC and resource dedicated to supporting the commander at that level. TCAEs or SIGINT elements at varying echelons draw information from each other in order to provide a more complete and detailed intelligence picture for their respective commanders.

ARMY TCAE (ATCAE)

8-7. The ATCAE, established at the national level, plays a significant role in TCAE operations by providing technical support oversight and providing access to national databases—in coordination with the NSA and other intelligence organizations. Additionally, the ATCAE works closely with NSA's IO element assisting in operations, monitoring technical capabilities, and providing liaison with Army's 1st Information Operations Command (Land). (This unit was previously called Land Information Warfare Activity.)

8-8. The ATCAE supports the Army's Quick Reaction Capability (QRC) SIGINT and signals research and target development (SRTD) operations. The QRC systems are responsive to the ground force commander's requirements and allow the ability to conduct SIGINT operations against modern communications systems.

REGIONAL TCAE (RTCAE)

8-9. The Army established RTCAEs to allow supported units at all echelons access to SIGINT regarding their respective AOI. The Army placed an RTCAE within each Regional Security Operations Center (RSOC). An RSOC is a joint service facility established by the Director, NSA, to conduct continuous security operations on selected targets in support of national and warfighter intelligence requirements using remoting technologies. The RSOC embeds a focus on the tactical commander's SIGINT requirements at a national level. It also creates a framework for enhanced interoperability among service SIGINT activities, especially within the context of a deployed JTF.

8-10. In addition to each RSOC, the RTCAE works closely with Theater TCAEs as well as Corps and Division SIGINT elements with the same regional focus. It provides analytic support and tailored products and answers RFIs. During peacetime and real-world operations RTCAEs facilitate intelligence reach support. They provide tailored technical support packages, support for surge and survey operations, analytic expertise, and signal data access and digital file transfers of live collection for Army tactical units.

THEATER TCAE

8-11. The Theater TCAE performs SIGINT and EW technical control and analysis and management. It provides SIGINT technical support for assigned, OPCON, and lower echelon SIGINT resources deployed in the theater. This includes mission tasking, processing, analyzing, and reporting of SIGINT data, information, and intelligence. The TCAE provides direction for the Theater Collection and Exploitation Battalion's SIGINT mission and for other theater tactical SIGINT assets.

SIGINT THEATER INTELLIGENCE BRIGADE/GROUP (TIB/TIG)

8-12. The TIB/TIG conducts operational level multidiscipline SIGINT operations. It provides timely intelligence to the commander throughout the full spectrum of operations. The TIB/TIG is structured to provide ground and aerial SIGINT support. It executes the full range of SIGINT missions or tasks.

SIGINT AT ECHELONS CORPS AND BELOW

8-13. There is a varying mixture of SIGINT assets within each corps. A reference listing the types of SIGINT assets found at echelons corps and below is listed in ST 2-50.

PLAN

8-14. An important SIGINT planning consideration is that if at all possible, SIGINT collection should be employed in conjunction with another intelligence discipline collection system. SIGINT is often used to cue, and be cued by, other ISR assets.

8-15. During planning, the G2/S2 should retrieve, update, or develop any required SIGINT databases. This includes effecting coordination with other SIGINT assets or elements that can support the operation.

8-16. SIGINT ground collection assets are usually placed in proximity to enemy signal sources due to the limited height of collector antennas, the low power output of threat or enemy emitters, and the line-of-sight (LOS) constraints imposed by terrain. Ground-based SIGINT teams are most effective when positioned to—

- Maximize threat emitter interception. This allows teams to overcome the constraints of threat emitter characteristics usually allowed by proximity to the threat.
- Minimize system receiver interference. This increases the potential capability of the team to acquire threat emitters in a timely manner.
- Optimize overlapping areas of intercept coverage. This ensures coverage of the AOIR and allows targets to be handed off from team to team if necessary.

8-17. Aerial SIGINT assets have additional planning requirements. Coordination must be effected with Army Airspace Command and Control (A2C2) for ingress and egress routes, restricted operating zones (ROZs) where the aircraft can travel on an intercept track, as well as with the supporting air operations center to determine the availability of asset types and times.

PREPARE

8-18. The G2/S2 ensures the SIGINT unit and asset leaders have effected all necessary coordination and conducted rehearsals. This includes establishing or verifying the operation of the SIGINT technical architecture. The G2/S2 also ensures all required SIGINT assets and resources are available, SIGINT reporting and dissemination channels and procedures are in place, and connectivity and interoperability exist with all pertinent SIGINT elements.

8-19. SIGINT OPCON is the authoritative direction of SIGINT activities, including tasking and allocation of effort, and the authoritative prescription of those uniform techniques and standards by which SIGINT information is collected, processed, and reported.

8-20. SIGINT operational tasking is the authoritative operational direction of and direct levying of SIGINT information needs by a military commander on designated SIGINT resources. These requirements are directive, regardless of other priorities, and are conditioned only by the capability of those resources to produce such information. Operational tasking includes authority to deploy all or part of the SIGINT resources for which SIGINT operational tasking authority (SOTA) has been delegated.

8-21. SOTA is the military commander's authority to operationally direct and levy SIGINT requirements on designated SIGINT resources; it includes authority to deploy and redeploy all or part of the SIGINT resources for which SOTA has been delegated.

COLLECT

8-22. SIGINT performs two major collection activities: signals intercept and direction finding (DF).

Signals Intercept

8-23. Signals intercept are those SIGINT actions used to search for, intercept, and identify threat electromagnetic signals for the purpose of immediate threat recognition. Signals intercept provides information required to answer PIRs, and other intelligence requirements in support of the ISR effort.

Direction Finding

8-24. Even when threat radio operators use COMSEC procedures, SIGINT teams can often intercept and approximate the location of the threat's signals. Specifically, SIGINT teams can use DF to determine—
- Movement of threat personnel or equipment.
- Locations of emitters associated with weapon systems and units.
- New emitter locations and confirm known emitter locations.
- Possible friendly targets the enemy intends to attack (lethal and non-lethal).

8-25. In addition to finding threat forces, DF operations can assist the (radio-equipped) friendly force by—
- Locating and vectoring assets or units during limited visibility.
- Locating downed aircraft and personnel radio beacons.
- Conducting signal security assessments.
- Locating sources of communication interference and jamming.

PROCESS

8-26. SIGINT processing involves converting intercepts of SIGINT into written and verbal reports, automated message, graphic displays, recordings, and other forms suitable for analysis and intelligence production. Since US forces routinely conduct operations against adversaries who speak languages other than English, SIGINT processing often also includes translation of these intercepts.

8-27. Due to the complexity of many SIGINT systems, automated processing may occur several times before SIGINT data or information receives any human interaction.

PRODUCE

8-28. The SIGINT producer must ensure the SIGINT product satisfies the associated intelligence requirements and that the product is in the required format. The quality, fidelity, and timeliness of SIGINT products are highly dependent upon the type of intercept, the collection system, the system's position in relation to the threat emitter, the weather (including space-based weather), as well as the SIGINT operator's ability to identify the appropriate threat signal activity.

8-29. SIGINT production results in some of the reports and formats mentioned in the process section above; however, the objective for SIGINT is to be used in an all-source analytical approach.

ANALYZE

8-30. The intelligence staff analyzes intelligence and information about the enemy's communications capabilities to determine appropriate SIGINT collection strategies. Conversely, a corresponding analysis of the friendly forces' SIGINT capabilities must be conducted to ensure the continued effectiveness of, or to improve upon, SIGINT collection.

8-31. SIGINT analysts also sort through large amounts of SIGINT and information and intelligence to identify and use only that which is pertinent to the CCIRs (PIRs and FFIRs).

DISSEMINATE

8-32. SIGINT of critical importance to the force, including answers to the CCIRs (PIRs and FFIRs), is disseminated via the most expeditious means possible. Due to the usually highly perishable nature of SIGINT, the most expeditious reporting means is often immediately augmented with a follow-up report or augmented by a report transmitted through additional means, enhancing the probability of receipt. Sometimes the most expeditious means of reporting critical SIGINT information to the commander is face to face.

8-33. For intelligence reach operations, SIGINT products are available and disseminated in a variety of forms: hardcopy, softcopy, direct viewing, or listening (television or radio). It is incumbent on the requestor to ensure that the SIGINT product can be transmitted over the available communications systems. This includes verifying the appropriate security level of the communications system.

ASSESS

8-34. The primary goal of the assess function when applied to SIGINT is to determine whether the results of SIGINT collection meet the requirements of the unit's ISR effort. SIGINT producers must assess all facets of SIGINT operations, from receipt of the ISR task to the dissemination of SIGINT, in an effort to determine their effectiveness. This assessment is not only directed at each SIGINT asset individually but also throughout the supporting SIGINT architecture and the unit's entire ISR effort.

8-35. The G2/S2 immediately assesses SIGINT products upon receipt for accuracy and relevance. He must inform the SIGINT producer of the extent to which the product answered the PIR or intelligence requirement. Providing feedback to the SIGINT producer—and collector—helps improve the effectiveness and efficiency of SIGINT.

Chapter 9
Measurement and Signatures Intelligence

DEFINITION

9-1. MASINT is technically derived intelligence that detects, locates, tracks, identifies, and/or describes the specific characteristics of fixed and dynamic target objects and sources. It also includes the additional advanced processing and exploitation of data derived from IMINT and SIGINT collection.

9-2. MASINT collection systems include but are not limited to radar, spectroradiometric, E-O, acoustic, RF, nuclear detection, and seismic sensors, as well as techniques for gathering NBC and other material samples.

9-3. It requires the translation of technical data into recognizable and useful target features and performance characteristics. Computer, communication, data, and display processing technologies now provide MASINT in support of commanders throughout the full spectrum of operations.

9-4. The subdisciplines within MASINT include, but are not limited to, the following:
- Radar Intelligence (RADINT). The active or passive collection of energy reflected from a target or object by LOS, bistatic, or over-the-horizon radar systems. RADINT collection provides information on radar cross-sections, tracking, precise spatial measurements of components, motion and radar reflectance, and absorption characteristics for dynamic targets and objectives. A SAR system, coupled with advanced MASINT processing techniques—
    - Provides a high resolution, day and night collection capability.
    - Can produce a variety of intelligence products that identify or provide change detection, terrain mapping, underwater obstacles, dynamic sensing of targets in clutter, and radar cross-section signature measurements.
- Frequency Intelligence. The collection, processing, and exploitation of electromagnetic emissions from a radio frequency weapon (RFW), an RFW precursor, or an RFW simulator; collateral signals from other weapons, weapon precursors, or weapon simulators (for example, electromagnetic pulse signals associated with nuclear bursts); and spurious or unintentional signals.
    - *Electromagnetic Pulses.* Measurable bursts of energy that result from a rapid change in a material or medium, resulting in an explosive force, produces RF emissions. The RF pulse emissions associated with nuclear testing, advanced technology devices, power and propulsion systems, or other impulsive events can be used to detect, locate, identify, characterize, and target threats.

- *Unintentional Radiation Intelligence (RINT)*. The integration and specialized application of MASINT techniques against unintentional radiation sources that are incidental to the RF propagation and operating characteristics of military and civil engines, power sources, weapons systems, electronic systems, machinery, equipment, or instruments. These techniques may be valuable in detecting, tracking, and monitoring a variety of activities of interest.
- **E-O Intelligence.** The collection, processing, exploitation, and analysis of emitted or reflected energy across the optical portion (ultraviolet, visible, and infrared) of the EMS. MASINT E-O provides detailed information on the radiant intensities, dynamic motion, spectral and spatial characteristics, and the materials composition of a target. E-O data collection has broad application to a variety of military, civil, economic, and environmental targets. E-O sensor devices include radiometers, spectrometers, non-literal imaging systems, lasers, or laser radar (LIDAR).
  - *Infrared Intelligence (IRINT)*. A subcategory of E-O that includes data collection across the infrared portion of the EMS where spectral and thermal properties are measured.
  - *LASER Intelligence (LASINT)*. Integration and specialized application of MASINT E-O and other collection to gather data on laser systems. The focus of the collection is on laser detection, laser threat warning, and precise measurement of the frequencies, power levels, wave propagation, determination of power source, and other technical and operating characteristics associated with laser systems—strategic and tactical weapons, range finders, and illuminators.
  - *Hyperspectral Imagery (HSI)*. A subcategory of E-O intelligence produced from reflected or emitted energy in the visible and near infrared spectrum used to improve target detection, discrimination, and recognition. HSI can detect specific types of foliage—supporting drug-crop identification; disturbed soil—supporting the identification of mass graves, minefields, caches, underground facilities or cut foliage; and variances in soil, foliage, and hydrologic features—often supporting NBC contaminant detection.
- **Spectroradiometric Products.** Include E-O spectral (frequency) and radiometric (energy) measurements. A spectral plot represents radiant intensity versus wavelength at an instant in time. The number of spectral bands in a sensor system determines the amount of detail that can be obtained about the source of the object being viewed. Sensor systems range from multispectral (2 to 100 bands) to hyperspectral (100 to 1,000 bands) to ultraspectral (1,000+ bands). More bands provide more discrete information, or greater resolution. The characteristic emission and absorption spectra serve to fingerprint or define the makeup of the feature that was observed. A radiometric plot represents the radiant intensity versus time. An example is the radiant intensity plot of a missile exhaust plume as the missile is in flight. The intensity or brightness of the object is a function of several conditions including its temperature, surface properties or material, and how fast it is moving. For each point along a time-intensity

radiometric plot, a spectral plot can be generated based on the number of spectral bands in the collector.
- Geophysical Intelligence. Geophysical MASINT involves phenomena transmitted through the earth (ground, water, atmosphere) and manmade structures including emitted or reflected sounds, pressure waves, vibrations, and magnetic field or ionosphere disturbances.
  - *Seismic Intelligence.* The passive collection and measurement of seismic waves or vibrations in the earth surface.
  - *Acoustic Intelligence.* The collection of passive or active emitted or reflected sounds, pressure waves or vibrations in the atmosphere (ACOUSTINT) or in the water (ACINT). ACINT systems detect, identify, and track ships and submarines operating in the ocean.
  - *Magnetic Intelligence.* The collection of detectable magnetic field anomalies in the earth's magnetic field (land and sea). An example is a Remotely Emplaced Battlefield Surveillance System (REMBASS) sensor detection indicating the presence and direction of travel of a ferrous object.
- Nuclear Intelligence (NUCINT). The information derived from nuclear radiation and other physical phenomena associated with nuclear weapons, reactors, processes, materials, devices, and facilities. Nuclear monitoring can be done remotely or during onsite inspections of nuclear facilities. Data exploitation results in characterization of nuclear weapons, reactors, and materials. A number of systems detect and monitor the world for nuclear explosions, as well as nuclear materials production.
- Materials Intelligence. The collection, processing, and analysis of gas, liquid, or solid samples. Materials intelligence is critical to collection against NBC warfare threats. It is also important to analyzing military and civil manufacturing activities, public health concerns, and environmental problems. Samples are both collected by automatic equipment, such as air samplers, and directly by humans. Samples, once collected, may be rapidly characterized or undergo extensive forensic laboratory analysis to determine the identity and characteristics of the sources of the samples.

ROLE

9-5. MASINT provides intelligence to the commander throughout the full spectrum of operations to facilitate situational understanding. MASINT can thwart many of the camouflage, concealment, and deception techniques currently used to deceive ISR systems.

9-6. MASINT is perceived as a "strategic" discipline with limited "tactical" support capabilities. But, by application of real-time analysis and dissemination, MASINT has a potential ability to provide real-time situation awareness and targeting not necessarily available to the classic disciplines. Specifically, MASINT "sensors" have unique capabilities to detect missile launch, detect and track aircraft, ships, and vehicles; do non-cooperative target identification (NCTI), combat assessment, and BDA; and detect and track fallout from nuclear detonations. Often, these contributions are the first indicators of hostile activities. For example, two EXOCET-equipped

Mirage F-1s were shot down during the Operation DESERT STORM due to MASINT collection and analysis. As evidenced by Operation IRAQI FREEDOM (OIF), MASINT will play a decisive role in the targeting of smart munitions with the signatures (fingerprint) of the targets they are seeking (for example, infrared signatures).

9-7. The MASINT systems most familiar on today's battlefield are employed by ground surveillance and NBC reconnaissance elements.

9-8. MASINT spans the entire EMS and its capabilities complement, rather than compete with, the other intelligence disciplines. MASINT provides, to varying degrees, the capability to—

- Use automatic target recognition (ATR) and aided target recognition (AiTR).
- Penetrate manmade and/or natural camouflage.
- Penetrate manmade and/or natural cover, including the ability to detect subterranean anomalies or targets.
- Counter stealth technology.
- Detect recently placed mines.
- Detect natural or manmade environmental disturbances in the earth's surface not discernible through other intelligence means.
- Provide signatures (target identification) to munitions and sensors.
- Enhance passive identification of friend or foe.
- Detect the presence of NBC agents to include prior to, during, or after employment.
- Detect signature anomalies that may affect target-sensing systems.

FUNDAMENTALS

9-9. Before discussing the functions of the intelligence process within a MASINT, the following paragraph provides an overview of organizational structure of MASINT.

9-10. Within DOD, the DIA provides central coordination for MASINT collection efforts through the Central MASINT Office. Each service, in turn, has a primary command or staff activity to develop requirements and coordinate MASINT effort. Army responsibility currently resides with INSCOM. Army weapons systems programs that require MASINT information to support system design or operations submit requests through the Army Reprogramming Analysis Team (ARAT) or INSCOM channels for data collection and processing. The S&TI community also performs MASINT collection and processing primarily to support R&D programs and signature development. Every S&TI center has some involvement in MASINT collection or production that reflects that center's overall mission (for example, NGIC does work on armored vehicles and artillery). Service R&D centers such as the Communications-Electronics Command (CECOM) Research, Development, and Engineering Center (RDEC), and Night Vision and Electronic Systems Laboratory are also involved in developing sensor systems for collecting and processing MASINT.

9-11. In addition to supporting the S&TI mission, INSCOM units also execute limited ground-based operational collection to support Theater and ASCC PIRs. This capability will expand upon the standup of INSCOM TIBs and TIGs.

PLAN

9-12. Some MASINT sensors can provide extremely specific information about detected targets, whereas other sensors may only be capable of providing an indication that an entity was detected. Additionally, there are varying capabilities of detection, identification, and classification among MASINT sensors. It is these varying capabilities that require synchronizing the employment of MASINT sensors both within the MASINT discipline and within the ISR effort as a whole. See FM 2-01 for more specific information on ISR synchronization.

9-13. As previously mentioned, there are many types of MASINT sensors. Depending on the type of sensor employed, a given MASINT collection target or NAI may not necessarily receive continuous coverage due to the possible conflict between the number and priority of targets and the number and availability of MASINT assets. However, a commander may decide to have continuous surveillance of certain targets by using his own MASINT assets (for example, REMBASS or Improved-REMBASS).

9-14. Another consideration when planning MASINT missions is whether to use active, passive, or a combination of both when planning MASINT coverage.

PREPARE

9-15. The G2/S2 MASINT related actions during the prepare function of the intelligence process include properly establishing or verifying the MASINT portion of the intelligence communications architecture functions. Additionally, the G2/S2 must ensure that required MASINT analytical assets and resources are prepared to provide support or are available through intelligence reach. Since the products of MASINT are not as well known as products from other intelligence disciplines, the G2/S2 must be aware of the types of MASINT products available to support the operation, and then educate the rest of his unit's staff on the use of these MASINT products. Lastly, the G2/S2 must also ensure MASINT reporting and dissemination channels and procedures are in place and rehearsals are conducted with all pertinent MASINT elements to ensure interoperability.

COLLECT

9-16. MASINT provides information required to answer PIRs and other intelligence requirements in support of the ISR effort. As stated earlier in this chapter, MASINT collection must not only be synchronized within its own discipline but also be synchronized and integrated into the unit's overall ISR effort in order to be effective.

9-17. MASINT sensors are employed throughout the full spectrum of operations from a variety of platforms—sub-surface, ground, marine, and aerospace.

PROCESS

9-18. Just as in the other intelligence disciplines, MASINT involves dealing with huge volumes of data that have to be processed before beginning analysis and production. The process function regarding MASINT involves converting esoteric data into a form that is suitable for performing analysis and producing intelligence. MASINT processing can include relatively simple actions such as converting a REMBASS sensor activation into a report, to a complex task such as processing HSI into a report identifying the composition and concentrations of carcinogenic chemicals present in the emissions from a factory upwind from a US forces encampment.

PRODUCE

9-19. Effective and timely MASINT requires personnel with diverse skill sets. The MASINT producer must ensure the MASINT product satisfies the associated intelligence requirements and that the product is in the required format. The quality, fidelity, and timeliness of MASINT products are highly dependent upon the type of target, the collection system, the system's position in relation to the target or NAI, and the weather, as well as the MASINT system operator's ability to identify the appropriate threat activity.

9-20. The objective of MASINT production is to be used in an all-source analytical approach.

ANALYZE

9-21. The intelligence staff analyzes intelligence and information about the enemy's equipment, doctrine, and TTP to determine appropriate MASINT collection strategies. Conversely, a corresponding analysis of the friendly force's MASINT capabilities must be conducted to ensure the continued effectiveness of, or to improve upon, MASINT collection.

DISSEMINATE

9-22. MASINT of critical importance to the force, including answers to the PIRs, is disseminated via the most expeditious means possible.

9-23. For intelligence reach operations, MASINT products are available and disseminated in a variety of forms. The requestor must ensure that the MASINT product can be transmitted over the available communications systems. This includes verifying the appropriate security level of the communications system.

ASSESS

9-24. The primary goal of the MASINT assess function is to determine whether the results of MASINT collection and production meet the requirements of the unit's ISR effort. MASINT producers must assess all facets of MASINT operations, from receipt of the ISR task to the dissemination of MASINT, in an effort to determine the effectiveness of MASINT. This assessment is not only directed at each MASINT asset individually but also throughout the supporting intelligence communications architecture, to include intelligence reach and the unit's entire ISR effort.

9-25. Also, the G2/S2 immediately assesses MASINT products upon receipt for accuracy and relevance. He must inform the MASINT producer of the extent to which the product answered the PIR or intelligence requirement. Providing feedback to the MASINT producer—and collector—helps improve the effectiveness and efficiency of MASINT.

Chapter 10

Technical Intelligence

DEFINITION

10-1. TECHINT is intelligence derived from the collection and analysis of threat and foreign military equipment and associated materiel.

ROLE

10-2. The strength of the US military lies, in part, to the diversity and extent of its technology base. While the US aspires to be the leader in integrating technology, the threat can achieve temporary technological advantage in certain areas by acquiring modern systems or capabilities. The world arms market is willing to provide these advanced systems to countries or individuals with the resources to pay for them. A concerted TECHINT program is vital to providing precise direction and purpose within the US R&D process to ensure quick and efficient neutralization of this advantage.

10-3. The role of TECHINT is to ensure that the warfighter understands the full technological capabilities of the threat. With this understanding, the warfighter can adopt appropriate countermeasures, operations, and tactics.

10-4. TECHINT has two goals within its role:

- To ensure the US armed forces maintain technological advantage against any adversary.
- To provide tailored, timely, and accurate TECHINT support to the warfighter throughout the entire range of military operations. This includes providing US forces intelligence, information, and training on foreign weapons systems to an extent that allows their use of CEE.

FUNDAMENTALS

10-5. The G2/S2 must understand how TECHINT assets are organized in order to properly apply the intelligence process, as the majority of TECHINT assets are located at EAC.

DEFENSE INTELLIGENCE AGENCY

10-6. DIA manages and reviews overall TECHINT activities. The S&TI Directorate within DIA is the action element for TECHINT. This directorate coordinates with external TECHINT agencies on non-policy matters concerning the production of S&TI. The following organizations provide TECHINT support under the control of DIA:

- **Armed Forces Medical Intelligence Center (AFMIC).** AFMIC, based at Fort Detrick, MD, is a DOD intelligence production center under DIA control. AFMIC is responsible for exploiting foreign medical materiel. The director supports the Army Foreign Materiel Exploi-tation

Program (FMEP) and Army medical R&D requirements. The director coordinates planning, programming, and budgeting with the Army Deputy Chief of Staff, Intelligence (DCS, G2).

- **Missile and Space Intelligence Center (MSIC).** MSIC, based at Redstone Arsenal, AL, is a DOD intelligence production center under DIA control and supports the FMEP. The MSIC acquires, produces, maintains, and disseminates S&TI pertaining to missile and space weapons systems, subsystems, components, and activities. The S&TI produced at MSIC also covers foreign state-of-the-art technology and research applicable to missiles.
- **Defense HUMINT Service.** DHS conducts worldwide HUMINT operations in support of foreign materiel acquisition (FMA) and foreign materiel exploitation (FME).

10-7. The organizations and agencies discussed below constitute the Army TECHINT structure.

ARMY DCS, G2

10-8. The Army DCS, G2 exercises general staff responsibility for all Army TECHINT activities. The Army DCS, G2 forms policies and procedures for S&TI activities, supervises and carries out the Army S&TI program, coordinates DA staff and MSC requirements for TECHINT, and is responsible for the Army Foreign Materiel Program (FMP).

INSCOM

10-9. Under the direction of Headquarters, Department of the Army (HQDA), INSCOM is responsible for peacetime TECHINT operations. HQ, INSCOM, fulfills its responsibilities through its TECHINT oversight function and manages the Army's Foreign Materiel for Training (FMT) Program and FMEP. It provides the interface with strategic S&TI agencies in support of FME and organizes, trains, and equips EAC TECHINT organizations during peacetime. TECHINT exploitation within INSCOM is performed by the following elements:

- **National Ground Intelligence Center.** HQ, INSCOM, exercises direct OPCON over the NGIC. NGIC produces and maintains intelligence on foreign scientific developments, ground force weapons systems, and associated technologies. NGIC analysis includes but is not limited to military communications electronics systems, types of aircraft used by foreign ground forces, NBC systems, and basic research in civilian technologies with possible military applications.
- **203d Military Intelligence Battalion.** The 203d MI Battalion is a multi-component unit headquartered at Aberdeen Proving Ground, MD, and is the Army's sole TECHINT battalion. It performs the following functions:
  - Conducts TECHINT collection and reporting in support of validated S&TI objectives.
  - Acts as the HQDA executive agent for foreign materiel used for training purposes.

- Conducts TECHINT training for DOD analysts and RC TECHINT personnel.
- Supports INSCOM's FMA and FME operations as directed.
- Analyzes and exploits foreign captured enemy documents (CEDs), equipment, weapon systems, and other war materiel.
- Reports on the capabilities and limitations of enemy combat materiel.
- Provides reports alerting the command to the tactical threat posed by technical advances in new or recently discovered foreign or enemy materiel.
- Provides countermeasures to any enemy technical advantage.
- Provides foreign or enemy equipment for troop familiarization and training.
- Provides recommendations on the reuse of CEM.
- Supervises evacuating items of TECHINT interest.
- Provides task-organized battlefield TECHINT teams to support a subordinate command's TECHINT effort.

US ARMY MATERIEL COMMAND (AMC)

10-10. AMC plays a significant support role in TECHINT. Among AMC elements are a series of RDECs, the Army Research Laboratory System, and the Test and Evaluation Command (TECOM). Each conducts highly technical evaluations of foreign equipment. In peacetime, the AMC conducts FME on equipment purchased by each laboratory and RDEC for the intelligence community and for DOD as part of the International Materiel Evaluation Program (IMEP). AMC elements include—

- The Foreign Ordnance Exploitation Team, which is located at the Fire Support Armaments Center (FSAC) in the Picatinny Arsenal. This team exploits foreign ground ordnance and develops render safe procedures (RSPs) for foreign ordnance. It also prepares detailed intelligence reports to support explosive ordnance disposal (EOD), MI, and US munitions developers.
- The Science and Technology Center Europe and the Science and Technology Center Far East, which have the responsibility of collecting information on foreign technical developments by attending arms shows and technology exhibitions.
- The Soldier Biological Chemical Defense Command, which is the headquarters for the US Army Technical Escort Unit (TEU). The mission of the TEU is to collect and escort (transport) chemical, biological, radiological, and nuclear (CBRN) samples for testing and evaluation.

10-11. There are many other agencies with TECHINT responsibilities within the DOD. Refer to FM 34-54 for more information on TECHINT.

PLAN

10-12. TECHINT collection usually begins when an organization or individual reports the recovery or acquisition of unusual, new, or newly employed threat materiel. However, there are often indications that the threat may be using materiel not yet associated with the threat among the myriad intelligence products available. Conversely, it may be known that the threat is using a particular item, the capabilities of which are unknown to US forces. It is in these cases that a unit may receive a TECHINT related ISR task. It is conceivable that such a task may be reflected in the PIR. An example of such a task is the identification or verification of suspected external modifications on a particular model of an enemy's main battle tank; the results of which are linked to a commander's decision.

10-13. TECHINT related ISR tasks should include a mission or target folder. At a minimum, this folder should include a description of the item, with its associated major combat systems, as well as handling instructions, reporting instructions, and a photograph or sketch of the item if available.

10-14. Additionally, the G2/S2 is responsible for ensuring the staff coordinates and establishes a plan for evacuating the desired materiel.

PREPARE

10-15. The G2/S2 must ensure that required TECHINT analytical assets, resources, and evacuation means are prepared to provide support. This includes verifying coordination effected with the task-organized battlefield TECHINT teams from the 203d MI Battalion. The G2/S2 must also ensure the means to report and disseminate TECHINT results to the unit and its soldiers are in place so that they can immediately adopt appropriate countermeasures, operations, or tactics in order to enhance their survival and mission accomplishment.

COLLECT

10-16. TECHINT collection includes capturing, reporting, and evacuating CEM. TECHINT collection begins when an organization or individual reports the recovery or acquisition of threat materiel. An item of materiel is exploited at each level, and continues on through succeeding higher levels until an appropriate countermeasure to neutralize the item's capabilities is identified or developed.

10-17. Army personnel (soldiers and civilian) and units will normally safeguard CEM and report it through intelligence channels to the first TECHINT element in the reporting chain. The location of this TECHINT element will be in accordance with the METT-TC factors; however, there will usually be TECHINT representation at the Corps G2 or the COCOM J2. The TECHINT representative or element will verify if the type of materiel is of intelligence value and determine its further disposition in conjunction with the unit's staff.

PROCESS

10-18. TECHINT processing starts (simultaneously with collection) with the capture of a piece of equipment of TECHINT value. This confirms that the enemy is indeed employing this materiel. In accordance with METT-TC factors, a TECHINT team may move to the location of the item at the capture site or wait until the item is evacuated before conducting a hasty exploitation. After hasty exploitation, the team decides if further processing is required. If it is, the items are sent to the first (or nearest) Captured Materiel Exploitation Center (CMEC). If the item is deemed to yield no immediate tactical intelligence value, it may still be evacuated to the S&TI centers in CONUS for further analysis if the systems represent a change in the technological posture of an enemy.

PRODUCE

10-19. Battlefield TECHINT teams normally report initial and secondary examinations of CEM using either a preliminary technical report or a complementary technical report.

- A preliminary technical report—
    - Includes a general description of the item reported and recommended RSP.
    - Alerts others to information that can be used immediately by tactical units.
- A complementary technical report is more in-depth and—
    - Follows a secondary or an in-depth initial examination.
    - Allows the CMEC to compare new information with intelligence holdings.

10-20. At each successive echelon of exploitation, TECHINT analysts add to the overall body of information on an item by either adding to previous reports or by preparing new reports. The CMEC or other national level S&TI activities prepare more advanced technical reports and analyses. These reports include—

- Detailed technical reports.
- Translation reports.
- Special technical reports.

10-21. Other TECHINT products include—

- CMEC publications such as operator manuals, maintenance manuals, TECHINT bulletins, and tactical user bulletins.
- S&TI analysis bulletins.
- Foreign materiel exploitation reports.

ANALYZE

10-22. TECHINT analysts use checklists established by S&TI agencies and the CMECs to analyze each type of the adversary's equipment for which requirements exist. Analysis always begins with what is, and what is not, known about the piece of equipment. TECHINT units maintain procedures and plans for sampling, analyzing, and handling materiel.

DISSEMINATE

10-23. TECHINT of critical importance to the force, including answers to the PIR, is disseminated via the most expeditious means possible.

10-24. Routine TECHINT reports and products are usually transmitted through the unit's existing intelligence communications architecture in the format of an intelligence information report (IIR) format. For intelligence reach operations, TECHINT products are available and disseminated in a variety of forms. The requestor must ensure that the TECHINT product can be transmitted over the available communications systems. This includes verifying the appropriate security level of the communications systems.

ASSESS

10-25. The primary goal of the TECHINT assess function is to determine whether the results of TECHINT production meet the unit's PIR or intelligence requirements. The G2/S2 immediately assesses TECHINT products upon receipt for accuracy and relevance. He must inform the TECHINT producer of the extent to which the product answered the PIR or intelligence requirement. Providing feedback to TECHINT analysts helps improve the effectiveness and efficiency of TECHINT.

10-26. The G2/S2 also assesses the success of the unit's ISR effort in accomplishing any TECHINT associated ISR task and shares his assessment with the staff and the pertinent units or personnel.

Chapter 11

Counterintelligence

DEFINITION

11-1. CI counters or neutralizes intelligence collection efforts through collection, CI investigations, operations, analysis and production, and functional and technical services. CI includes all actions taken to detect, identify, exploit, and neutralize the multidiscipline intelligence activities of friends, competitors, opponents, adversaries, and enemies. It is the key intelligence community contributor to protect US interests and equities. Figure 11-1 shows the CI overview.

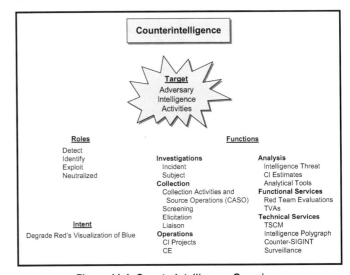

Figure 11-1. Counterintelligence Overview.

ROLE

11-2. The role of CI is to detect, identify, exploit, and neutralize all adversary intelligence entities targeting US and multinational interests. CI will focus on countering adversary intelligence collection activities targeting information or material concerning US personnel, activities, operations, plans, equipment, facilities, publications, technology, or documents—either classified or unclassified—without official consent of designated US release authorities, for any purpose that could cause damage or otherwise adversely impact the interests of national security of the US ability to fulfill national policy and objectives. Adversary intelligence threats include but are not limited to any (US, multinational, friendly, competitor, opponent, adversary, or recognized enemy) government or NGOs, companies, businesses, corporations, consortiums, groups, agencies, cells or persons, terrorists, insurgents, guerrilla entities, and persons whose demonstrated actions, views, or opinions are inimical to US interests.

11-3. CI elements are instrumental in contributing to situational awareness in the AOI. CI elements may corroborate other intelligence discipline information as well as cue other intelligence assets through the CI core competencies and through CI technical services. The CI core competencies are collection, investigations of national security crimes within the purview of CI, operations, and analysis and production. CI technical services include computer network operations, technical surveillance countermeasures (TSCM), and polygraph. CI focuses on combating adversary intelligence activities targeting Army personnel, plans, operations, activities, technologies, and other critical information and infrastructure.

COUNTERINTELLIGENCE FUNCTIONS

11-4. CI functions are interrelated, mutually supporting, and can be derived from one another. No single function or technical capability can defeat adversary intelligence efforts to target US interests. CI functions are discussed below.

CI INVESTIGATIONS

11-5. Investigative activity is essential to countering the adversary intelligence threat to Army interests. CI places emphasis on investigative activity to support force, infrastructure and technology protection, homeland defense, information assurance, and security programs. CI investigations focus on resolving allegations of known or suspected acts that may constitute National Security Crimes under US law. The primary objective in any CI investigation is the detection, identification, exploitation, and/or neutralization of adversary intelligence threats directed against the US Army. CI investigations are also conducted to identify systemic security problems that may have damaging repercussions to Army operations and national security interests. All CI investigations are conducted within guidelines established in AR 381-10, AR 381-12, AR 381-20, applicable DOD policy and directives, and US laws.

CI OPERATIONS

11-6. CI operations are characterized as those activities that are not solely associated with investigative, collection, analysis, or production functions. CI operations can be either offensive or defensive in nature; they are derived from or transition to a collection or investigative activity depending on the scope, objective, or continued possibility for operational exploitation. CI operations fall into two categories: CI support operations and CI sensitive operations.

- **CI Support Operations.** These are defensive operations used to support ARFOR and technology protection, security projects, and programs. They include technical services support, support to acquisition, FP, special access, international security, foreign visitor or contact, treaty verification, information assurance, homeland defense, and other approved projects and programs.
- **CI Sensitive Operations.** These operations are generally offensive in nature and involve direct or indirect operations against a known adversary intelligence threat. These operations include counter-espionage (CE) and CI projects and are conducted by designated units.

CI COLLECTION

11-7. **Collection Activities.** CI elements conduct collection activities focused on adversary intelligence threats that target US and multinational interests. CI collection is conducted through the use of sources, assets, official contacts, and other human or multimedia sources to obtain information that impacts the supported unit. CI will not be used as a substitute for Army HUMINT collection. These activities are designed to collect specific information or develop leads concerning adversary intelligence collection requirements, capabilities, efforts, operations, structure, personalities, and methods of operation targeting US and multinational interests. CI collection can result from ongoing CI investigations or operations or serve to initiate CI investigations and/or operations.

11-8. **Liaison.** CI elements conduct liaison with US, multinational, and HN military and civilian agencies, to include NGO, for the purpose of obtaining information of CI interest and coordinating or deconflicting CI activities. Liaison activities are designed to ensure a cooperative operating environment for CI elements and to develop CI leads for further exploitation.

11-9. **CI Collection Activities and Source Operations (CASO).** CASO is used to collect information on direct threats to US Army installations, organizations, activities, and personnel. The CASO program is not intended to be used as a substitute for tactical HUMINT contact operations. CASO can be used to initiate CI investigations, identify potential leads for offensive operations, or develop additional CASO leads. Only designated units, in accordance with applicable policy, will be authorized to pursue investigative and offensive operational leads from ongoing CASO operations.

11-10. **Screening.** CI Special Agents work jointly with HUMINT Collectors during screening operations to identify civilians on the battlefield, EPWs, detainees, and other noncombatants who may have information of CI interest or to develop CI leads. CI screening is also conducted during the process of hiring HN citizens for Army and DOD employment. Information obtained during screening operations may be used to initiate CI investigations and operations or to cue other intelligence collection disciplines such as HUMINT, IMINT, SIGINT, and MASINT.

11-11. **Debriefing.** CI Special Agents conduct debriefings of friendly force, HN, or the local population who may have information of CI interest regarding adversary intelligence collection or targeting efforts focused on US and multinational interests.

11-12. **Functional Services.** CI elements conduct functional services to assist traditional CI activities of investigations, collection, operations, analysis, and production; they also provide tailored support to US, DOD, and Army protection and security programs for commanders at all echelons. CI elements may use one or several of the functional services simultaneously to provide tailored support to a particular CI mission or supported program. CI functional services consist of—

- CI threat vulnerability assessments (TVAs).
- Adversary intelligence simulation (Red Team Evaluation).
- Covering agent support.

11-13. **CI Technical Services.** CI technical services are used to assist the CI functions of investigations, collections, and operations or to support functional services conducted by CI elements. For additional information on CI technical services, refer to AR 381-20 (SECRET/NOFORN) and FM 34-5. CI technical services consist of—

- Surveillance.
- Intelligence polygraphs.
- TSCM.
- Computer Network Operations (CNO).
- IO.
- Counter-Signals Intelligence (C-SIGINT).

ANALYSIS

11-14. IPB, all-source, and single-source analysis are used to template adversary intelligence activities. Analysis is also used to make recommendations to the commander on how to counter adversary intelligence efforts and to refine CI activities to potentially neutralize and/or exploit those efforts and to continually focus the efforts of CI teams. CI analysis focuses on the multidiscipline adversary intelligence collection threat targeting information on US multinational personnel, operations, activities, technology, and intentions. Raw information, open-source material, and finished intelligence products are analyzed in response to local and national requirements. Analysis occurs at all levels from tactical to strategic.

- At the tactical level, CI teams focus their efforts on supporting mission requirements and contributing to the all-source COP.
- Operational analysis is used to assess how adversary intelligence views and targets US interests; identify US vulnerabilities that could be exploited or targeted; and to determine adversary intelligence targeting methods of operation (MOs).
- CI collection priorities; assessing adversary intelligence technical options for countering US weapons and intelligence systems; and

assessing the impact of technology transfer activities on the US military's technological overmatch.

PRODUCTION

11-15. CI products consist of, but are not limited to, target nomination, CI input to TVAs, CI estimates, and investigative and intelligence information reports. Finalized intelligence derived from CI activities are incorporated into joint and national intelligence databases, assessments, and analysis products. CI products are also incorporated into the COP to support battlefield situational awareness. CI production takes place at all levels.

- Operational and tactical production includes tactical alerts, spot reports, and current intelligence; CI threat and/or vulnerability assessments tailored to specific activities, units, installations, programs, or geographic areas; CI studies to support contingency planning and major exercises; studies of adversary intelligence organization, MO, personnel, activities, and intentions that pose a current or potential threat to the supported command.
- Strategic products include assessments supporting national and Army programs including SAPs and acquisition programs; worldwide assessments of the organization, location, funding, training, operations capabilities, and intentions of terrorist organizations; global trends in adversary intelligence MO; after-action studies of individual espionage cases; analyses of the intelligence collection capabilities of international narcotics trafficking organizations; and multimedia threat products to support Army CI awareness programs.

TECHNICAL SERVICES

11-16. CI organizations with technically trained CI Special Agents are chartered with providing unique technical capabilities to augment CI investigations, collection, and operations. These technical capabilities are not used as substitutes for CI activities, but support traditional CI techniques employed to counter and neutralize adversary intelligence activities targeting US interests.

11-17. Selected CI Special Agents are trained to identify human deception indicators during the course of investigations, operations, and collection missions. Required training and skills include the use of polygraph and emerging biometric technologies that can recognize indications of deception by human sources, contacts, and subjects of investigations, as well as analysis and reporting of results.

11-18. TSCM CI Special Agents are trained in sophisticated electronic and sensing equipment to identify technical collection activities carried out by adversary intelligence entities. The use of TSCM is critical to ensuring that sensitive and/or restricted areas are clear of any adversary intelligence-placed active or passive electronic sensing, or eavesdropping or collection devices. These areas include SCIF, secure working and planning areas, and C2 facilities.

11-19. CNO consist of computer network attack (CNA), computer network defense (CND), and computer network exploitation (CNE). CI Special Agents are specially trained in the areas of computer operation, network theory and administration, and forensics, along with IO in order to ensure US information dominance. The reliance on networked systems will result in greater emphasis being placed on information assurance. Specially selected CI agents will be trained in CNO in order to assist in protecting US information and information systems while exploiting and/or attacking adversary information and information systems.

11-20. For more information on technical services, refer AR 381-20 and FM 34-5.

OPERATIONAL EMPLOYMENT

11-21. Army CI supports the full spectrum of military operations. The Army requires a well-trained CI force consisting of AC, RC, civilian government employee, and contractor personnel. CI elements are focused on and dedicated to detecting, identifying, neutralizing, and/or exploiting adversary intelligence elements attempting to collect information on US forces. Effective employment of Army CI elements in all phases of operations and at all levels from tactical to strategic is paramount to countering any adversary intelligence threat to US interests and resources. Moreover, Army CI protects special programs that provide R&D, acquisition, and integration of technologies leading to future technological overmatch. CI elements have the capability and authority to conduct complex, nontraditional operations in this Information Age to CE, to both protect critical technologies and satisfy Army, DOD, and national level CI objectives.

LEVELS OF EMPLOYMENT

Strategic and Departmental

11-22. Strategic and departmental operations will be conducted by CI elements supporting national, DOD, and DA required missions (for example, support to NATO and special operations and missions). Strategic and departmental CI will conduct compartmented investigations and operations to affect the knowledge of adversary intelligence regarding CONOPS and defense information. Army CI will execute the full range of CI functions and missions at the strategic and departmental level including CI investigations and operations, CE, technology protection, SAP support, treaty verification, and technical CI services (polygraph, TSCM, and computer forensics). Strategic and departmental CI will also support SOF and special mission units (SMUs) within the scope of applicable national, DOD, and DA security policies and regulations.

Operational

11-23. Operational missions of CI elements will support combatant commanders, generally in geographic AOIR. Operational CI elements will focus on threat identification and countering regional adversary intelligence threats. Operational level CI activities and functions include CI investigations and operations, CE, technology protection, SAP support, treaty verification, and technical CI services (polygraph, TSCM, and computer forensics). CI elements must be capable of quickly transitioning from a peacetime mission to crisis operations to support combatant commander requirements. Theater CI assets will conduct unilateral to multinational operations in designated theaters. Operational elements may also be deployed to support or reinforce tactical forces in CONOPS.

Tactical

11-24. CI teams will conduct operations throughout the battlespace during CONOPS. CONOPS support activities include conduct of CASO, limited CI investigative capability, personnel security investigations, screenings, and debriefings. CI activities in CONOPS focus on countering the adversary intelligence threat and assisting in conducting TVAs. TVAs will be conducted in conjunction with MPs, Engineers, and Medical Service personnel to provide the commander with a comprehensive FP assessment. During peacetime, organic tactical CI teams conduct activities in accordance with approved regulations and command guidance.

SUPPORT TO CONTINGENCY OPERATIONS

11-25. The initial phase of operations from PME to MTW lays the foundation of future team operations. In general, the priority of effort focuses inward on security of operating bases, areas of troop concentration, and C2 nodes to identify the collection threat to US forces that could be used by adversary elements to plan hostile acts against US activities and locations.

11-26. Once security of the operating bases has been established, the operational focus of CI teams shifts outside the operating base to continue to detect, identify, and neutralize the collection threat to US forces as well as to provide I&W of hostile acts targeting US activities. The CI team uses several collection methods, to include CASO, elicitation, and liaison, to answer the supported commander's requirements. This is referred to as the continuation phase. The CI team conducts CI investigations to identify, neutralize, and exploit reported threat intelligence collection efforts.

11-27. A key element to the CI team's success is the opportunity to spot, assess, and develop relationships with potential sources of information. Operating as independent teams, without being tied to ISR or combat assets, enables the CI team's maximum interaction with the local population, thereby maximizing the pool of potential sources of information. Along with the opportunity to spot, assess, and interact with potential sources of information, a second key element of a CI team's success is its approachability to the local population. A soft posture enables a CI team to appear as non-threatening as possible. Experience has shown that the local population in general is apprehensive of fully and openly armed patrols and soldiers moving around population centers.

11-28. During some operations, civilian attire or nontactical vehicles may be used to lower the CI team profile. In some special situations, these measures are taken to make the operation less visible to the casual observer. Also, in some cultures, sharing food and beverages among friends is expected; exceptions to restrictions or general orders should be considered to facilitate successful CI team operations, many of which are geared towards developing relationships with potential sources of information.

SUPPORT TO INSTALLATIONS AND OPERATING BASES

11-29. CI teams, as part of a multi-agency team consisting of MPs, CA, medical, and EOD, support the conduct of TVAs of installations and operating bases to identify the intelligence threat to the operating locations. Detailed TVAs identify weaknesses in operational and physical security procedures and recommend countermeasures to mitigate intelligence collection on friendly forces limiting the ability to plan hostile acts on US activities and locations. CI activities supporting installations and operating bases include—

- Interviewing walk-in sources and locally employed personnel.
- Screening local national (LN) hires. Commanders, staff planners, and SIOs should always provide input to personnel assigned to establish and negotiate contracts using local national (LN) hires. This requirement ensures that LN hires can be screened, interviewed, and in some instances used as CI sources or assets in order to provide intelligence information that impacts the security of the base camp.
- Debriefing friendly force personnel who are in contact with the local population, such as—
  - ISR patrols.
  - MP patrols.
  - Combat patrols.
  - Liaison personnel.
  - CA and PSYOP teams.
- Conducting limited local open-source information collection.
- Providing support to TVAs of the base camp.

TACTICS, TECHNIQUES, AND PROCEDURES

11-30. At the CI team level, team members conduct mission analysis and planning specific to their AO. Backwards planning and source profiling are used extensively to choose CI targets. To verify adequate area coverage, the CI team may periodically develop and use CI target overlays and other CI analytical tools that illustrate the CI situation, identify CI gaps, and help refocus the collection effort.

11-31. The CI team is also in constant contact with the supported S2 and the other ISR assets (Scouts, PSYOP, CA, and MP) in order to coordinate and deconflict operations and to cross-check collected information. The supported unit S2, with the help of the CI team, regularly and systematically debriefs all ISR assets.

11-32. The CI team must be integrated into the supported unit's ISR plan. The CI OMT chief will advise the supported unit on the specific capabilities and requirements of the team to maximize mission success.

OPERATIONAL RISK MITIGATION

11-33. The employment of CI teams includes varying degrees of contact with the local population. As the degree of contact with the population increases, both the quantity and quality of CI collection increases. In many instances, however, there is a risk to the CI team inherent with increased exposure to the local population. The decision at what level to employ a CI team is METT-TC dependent. The risk to the CI assets must be balanced with the need to collect priority information and to protect the force as a whole. ROE, SOFA, direction from higher headquarters, and the overall threat level may also restrict the deployment and use of CI teams. The commander should consider exceptions to the ROE to facilitate CI collection.

11-34. Risks are minimized through the situational awareness of CI team members. They plan and rehearse to readily react to any situation and carry the necessary firepower to disengage from difficult situations. If it becomes necessary to call for assistance, adequate and redundant communications equipment is critical. These scenarios and actions should be trained prior to deployment into a contingency area and rehearsed continuously throughout the deployment.

11-35. A supported unit commander is often tempted to keep the CI team "inside the wire" when the THREATCON level increases. The supported commander must weigh the risk versus potential information gain when establishing operational parameters of supporting CI teams. This is necessary especially during high THREATCON levels when the supported unit commander needs as complete a picture as possible of the threat arrayed against US and multinational forces.

11-36. When it is not expedient to deploy the CI team independently due to threat levels or other restrictions, the team can be integrated into other ongoing operations. The CI team may be employed as part of a combat, ISR, or MP patrol or used to support CA, PSYOP, engineer, or other operations. This method reduces the risk to the team while allowing a limited ability to collect information. It has the advantage of placing the team in contact with the local population and allowing it to spot, assess, and interact with potential sources of information. However, this deployment method restricts collection by subordinating the team's efforts to the requirements, locations, and timetables of the unit or operation into which it is integrated and does not allow for the conduct of sensitive source operations. **This method of employment should be considered a last resort.**

COUNTERINTELLIGENCE EQUIPMENT

11-37. Basic C2, transportation, and weapons requirements do not differ significantly from most soldier requirements and are available as unit issue items. However, CI teams have unique communications, collection, processing, and mission-specific requirements.

COMMUNICATIONS

11-38. **Dedicated and Secure Long-Range Communications.** These are keys to the success of the CI team mission. CI team operations require a secure, three-tiered communications architecture consisting of inter/intra-team radios, vehicle-based communications, and a CI and HUMINT base station.

11-39. **Communications Network.** The CI team must have access to existing communications networks such as the tactical LAN. The CI team must also be equipped with its own COMSEC devices. It is imperative that the CI team acquire access to the public communication system of the HN. This can be in the form of either landlines or cellular telephones. Such access enables the CI team to develop leads which can provide early indicators to US forces.

11-40. **Interoperability.** Communications systems must be equipped with an open-ended architecture to allow for expansion and compatibility with other service elements, government organizations, NGOs, and multinational elements to effectively communicate during CONOPS. All ISR systems must be vertically and horizontally integrated to be compatible across all BOSs and with Legacy and Interim Force elements.

11-41. **SOTM.** To provide real-time and NRT information reporting, CI elements must have the capability to transmit voice, data, imagery, and video while on the move. CI teams must be able to transmit while geographically separated from their parent unit while operating remotely. This broadband requirement can only be achieved through a SATCOM capability and must be achievable while mobile.

CI COLLECTION AND PROCESSING SYSTEMS

11-42. The CI team must rely on automation to achieve and maintain information dominance in a given operation. With time, effective collection planning and management at all echelons, the CI team can collect a wealth of information. The sorting and analysis of this information in a timely and efficient manner is crucial to operations. Automation helps the CI team to report, database, analyze, and evaluate the collected information quickly and to provide the supported unit with accurate data in the form of timely, relevant, accurate, and predictive intelligence.

11-43. Automation hardware and software must be user friendly as well as interoperable among different echelons and services. They must interface with the communications equipment of the CI team as well as facilitate the interface of audiovisual devices. Technical support for hardware and software must be available and responsive.

11-44. The demand for accurate and timely CI reporting, DOCEX, and open-source information has grown tremendously. Biometric (physiological, neurological, thermal analysis, facial and fingerprint recognition) technologies will allow rapid identification, coding, and tracking of adversaries and human sources; as well as cataloging of information concerning EPWs, detainees, and civilians of CI interest on the battlefield. Biometrics will also provide secure authentication of individuals seeking network or facility access.

11-45. CI teams work with multinational forces and other foreign nationals and require the ability to communicate in their respective languages. Often CI personnel have little or no training in the target language, and lack of skilled interpreters can hinder CI activities. CI teams require textual and voice translation devices, source verification, and deception detection machines (biometrics) to improve collection capability and accuracy.

11-46. CI teams require dynamic MLT tools that provide both non-linguists and those with limited linguist skills a comprehensive, accurate means to conduct initial CI screenings and basic interviews in a variety of situations. CI elements will focus on in-depth interviews and communications with persons of higher priority. MLT tools minimize reliance on contract linguists and allow soldiers to concentrate on mission accomplishment.

MISSION SPECIFIC

11-47. The CI team may conduct night operations and must be equipped with NVDs for its members, and photographic and weapons systems. The CI team also may operate in urban and rural areas, where the threat level can vary from semi-hostile to hostile. The safety of the CI team can be enhanced with equipment that can detect, locate, suppress, illuminate, and designate hostile optical and E-O devices. In addition, high power, gyro-stabilized binoculars, which can be used from a moving vehicle, increases the survivability of the CI team and also gives the team another surveillance and collection device.

11-48. Some of the CI team missions may require the documentation of incidents. The CI teams can use the following equipment in their open-source collection efforts.

- Small, rugged, battery-operated digital camcorders and cameras which are able to interface with the collection and processing systems as well as communication devices.
- GPSs that can be mounted and dismounted to move in the AO efficiently.
- Short-range multichannel RF scanning devices that can also identify frequencies which enhance their security.
- In some cases CI teams require a stand-off, high resolution optical surveillance and recording capability that can provide target identification at extended ranges to protect the intelligence collector while avoiding detection by the adversary target. An advanced optical capability provides intelligence collectors the ability to locate and track adversary targets (passive and hostile) for identification, collection, and target exploitations.

INTEGRATION OF LINGUISTS

11-49. Integrating linguists into the CI team should take place as soon as possible. Security clearances and contractual agreements will help the team determine the level of integration.

11-50. Along with the basic briefing of what is expected of the civilian linguists as interpreters, CI teams should be informed about the civilians' chain of command and the scope of their duties beyond interpreting. The CI team leader must ensure that linguists are trained and capable of completing all tasks expected of them.

BATTLE HAND-OFF

11-51. CI teams are always engaged. A good battle hand-off is critical to smooth transition and mission success. The battle hand-off can directly contribute to mission success or failure of the outgoing team, but especially of the incoming team. The battle hand-off begins the first day the CI team begins to operate in an AO. Regardless of how long the team believes it will operate within the AO, it must ensure there is a seamless transition to an incoming team, other US unit, or agency. The CI team accomplishes this transition by establishing procedures for source administration, database maintenance, and report files.

11-52. Teams must plan and implement a logical and systematic sequence of tasks that enables an incoming team to assume the operations in the AO. Adequate time must be allotted for an effective battle hand-off. In some environments, a few weeks may be necessary to accomplish an effective battle hand-off. Introductions to sources of information, especially CASO sources, are critical, and teams must prioritize their time. During this time the outgoing CI team must familiarize the new CI team with all aspects of the operation, which include past, present, and planned activities within the AO. Area orientation is critical. These include major routes, population centers, potential hot spots, and other points of interest (such as police stations, political centers, and social centers).

ORGANIZATION

11-53. CI activities require a complex C2 relationship to ensure that the requirements of the supported commander are fulfilled while balancing the need for strict integrity and legality of CI operations. This complex relationship balances the role of the SIO as the requirements manager and the 2X as the mission manager with the MI commander as the asset manager.

COMMAND VERSUS CONTROL

11-54. ARFOR will normally deploy as part of a joint, multinational, and/or combined operation. In all cases, commanders at each echelon will exercise command over the forces assigned to their organization. Command includes the authority and responsibility for effectively using resources, planning for and employment of forces, and ensuring that forces accomplish assigned missions. Leaders and staffs exercise control to facilitate mission accomplishment.

11-55. While the MI commander supervises subordinates and produces reports, the *2X synchronizes activities between intelligence units and provides single-source processing and limited analysis. (*2X " indicates 2X functions at all levels.) While the MI commander takes care of the operators executing missions, the *2X obtains the data and reports from higher echelons required to execute the missions.

STAFF RESPONSIBILITIES AND FUNCTIONS

11-56. The *2X staff is responsible for the integration, correlation, and fusion of all Human Sensor information into the Intelligence BOS within the *2X AOIR. The *2X is also responsible for analyzing adversary intelligence collection, terrorist and sabotage activities, developing countermeasures to defeat threat collection activities, identifying and submitting collection requirements to fill CI collection gaps, and providing input to the all-source picture regarding adversary intelligence activities.

11-57. The *2X Staff Officer provides CI and HUMINT collection expertise. The *2X—

- Is the single focal point for all matters associated with CI and HUMINT in the AOIR.
- Is the CI and HUMINT advisor to the G2 and commander.
- Is an extension of the collection manager and ensures that the best asset or combinations of assets are used to satisfy information requirements.
- Along with his subordinate elements—CICA, HOC, OSC, CIAC, and HAC—exercises technical control over his assigned Army CI and HUMINT elements in the designated AOIR.
- Is the principal representative of the G2 and the commander when coordinating and deconflicting CI and HUMINT activities with national or theater agencies operating in the AOIR.
- Supports specific RM efforts in conjunction with the requirements manager through the planning and coordination of CI and HUMINT operations; the review and validation of CI requirements; the recommendation for assignment of tasks to specific collectors; and the conduct of liaison with non-organic HUMINT collection. This liaison includes national level and multinational force assets for source deconfliction and special activities outside the *2X AOIR.
- Will provide OMTs with capability to reach back to current database information, technical information and guidance, and source deconfliction necessary to monitor the collection activities of the CI teams.
  - *CICA.* The CICA is responsible for coordinating and synchronizing all CI activities in the designated AOIR. The CICA exercises technical control over all CI entities in the designated AOIR and deconflicts CI activities with higher, lower, and adjacent CI elements. The CICA accomplishes all responsibilities through coordination with the operational units and other *2X staff elements.
  - *OSC.* The OSC in the *2X staff maintains the source registry for all CI activities in the designated AOIR. The OSC provides management of intelligence property book operations, source incentive programs, and ICFs.
  - *CIAC.* The CIAC analyzes adversary intelligence collection capabilities. The CIAC leverages all intelligence discipline reporting and analysis to counter threat collection capabilities against the deployed force. CIAC analysis provides information and analysis to the COP.

11-58. The ACE and JISE CI analysis team analyzes threat intelligence collection and the intelligence collection efforts of foreign organizations involved in terrorism and sabotage in order to develop countermeasures against them. CI analysis cross-cues IMINT, SIGINT, MASINT, and TECHINT resources in addition to CI-related HUMINT reporting and analysis to counter threat collection capabilities against the deployed force. While the HAC supports the positive collection efforts of the force, the CI analysis team supports the "defend" aspects of the commander's FP program.

11-59. CI analysis is the analysis of the adversary's HUMINT, IMINT, SIGINT, and MASINT capabilities in support of intelligence collection, terrorism, and sabotage in order to develop countermeasures against them. It involves a reverse IPB process in which the analyst looks at US forces and operations from the threat's perspective. CI analytical products are an important tool in the COA development in the MDMP. This analytical tool supports the commander's FP program and facilitates the nomination of CI targets for neutralization or exploitation. (See FM 2-01.2 (FM 34-60) for more information on CI analysis.) Specifically, CI analysis—

- Produces and disseminates CI products and provides input to INTSUMs.
- Provides collection requirements input to the CICA.
- Analyzes source reliability and credibility as reflected in reporting and communicating that analysis to the collector.
- Nominates CI targets for neutralization or exploitation.
- Identifies and submits CI-related requirements to fill collection gaps.
- Assists HAC personnel in focusing the CI aspects of the HUMINT collection program.
- Presents CI analysis products such as CI estimates, target lists, reports, and graphics that support the commander.

11-60. For intelligence reach operations, CI products are available and disseminated in a variety of forms. It is incumbent on the requestor to ensure the CI product can be transmitted over the available communications systems. This includes verifying the appropriate security level of the communications systems.

CI TEAM STRUCTURE

11-61. **OMT.** The OMT is a four-person team consisting of a WO, two NCOs, and a junior enlisted soldier. (Civilians may be inserted into this structure as appropriate.) Rank structure and standards of grade for OMTs will vary depending upon the skill sets required and mission focus. CI OMTs will provide operational guidance for 1 to 4 CI teams, depending on mission focus and operational tempo. When two or more CI teams are deployed in a DS role, an OMT is also deployed to provide technical control. The OMT works closely with the supported S2 and ACT to furnish current threat information and to answer the supported commander's PIRs and IRs. OMTs coordinate with the supported 2X and manage subordinate CI teams to—

- Provide guidance and technical control of operational activity.
- Provide the collection and operational focus for CI teams.

- Provide quality control and dissemination of reports for subordinate CI teams.
- Conduct single-discipline CI analysis and assist in mission analysis for the supported commander.
- Act as a conduit between subordinate CI teams, the CICA, and supported unit headquarters.
- Provide administrative support for subordinate CI teams to include reporting mission and equipment status to the CICA and the supported unit headquarters.
- Educate the supported commander on the capabilities of the CI teams.
- Integrate the CI teams directly into the maneuver commander's ISR planning.

11-62. **CI Team.** The CI team is a four-person team consisting of two NCOs and two junior enlisted personnel. Rank structure and standards of grade for CI teams will vary depending upon the skill sets required and mission focus. CI teams are trained to execute the full range of CI functions; however, they may be assigned to mission-focused elements (for example, CE, CI projects). Assignment to a TSCM, polygraph, or information warfare team requires additional, specialized technical training.

Appendix A
Intelligence and Information Operations

THE INFORMATION ENVIRONMENT

A-1. The information environment is the aggregate of individuals, organizations, and systems that collect, process, store, display, and disseminate information; also included is the information itself. (FM 3-0) The information environment, it should be noted, is not an exclusively military one; in fact, the military applications of information are almost obscured in today's universal usage of the information spectrum by national, international, and non-state players.

THE COMMANDER AND INFORMATION

A-2. Information is facts, data, or instructions in any medium or form; the meaning that a human assigns to data by means of the known conventions used in their representation. (JP 1-02) Information provides the key to battlefield success in the 21st century. Commanders must have detailed information to command. Information is the medium that allows the commander's decisionmaking and execution cycle to function. Information gives direction to actions by the force, identifies the enemy's centers of gravity, provides COAs for force activity, and enables the force to accomplish its operational mission.

INFORMATION SUPERIORITY

A-3. Information superiority is the operational advantage derived from the ability to collect, process and disseminate an uninterrupted flow of information while exploiting or denying an adversary's ability to do the same. (FM 3-0) Relevant information drawn from intelligence supports the creation and development of situational understanding that contributes directly to information superiority during decisive operations. The requirement for information superiority is not new; what is new is that today's information technologies are creating a base of knowledge for military planning and execution that is unprecedented in scope, volume, accuracy, and timeliness. This means commanders receive accurate, timely information that enables them to make better decisions and act faster than their adversaries.

A-4. Information superiority, however, is neither a staple in today's battlespace nor is it necessary in military operations as a constant condition; rather, at specific times during operations, information superiority becomes a key enabler for assuring military success. At the operational level of predominant interest to the land component commander (LCC), information superiority is realized through the integration of operational level, interdependent ISR, information management, and IO to gain and maintain operational initiative and to achieve an operational advantage.

INFORMATION OPERATIONS

A-5. IO are the employment of core capabilities of electronic warfare, computer network operations, PSYOP, military deception, and operations security, in concert with specified supporting and related capabilities to affect or defend information and information systems, and to influence decisionmaking.... Information systems are...the equipment and facilities that collect, process, store, display, disseminate information. These include computers, hardware, software, and communications, as well as policies and procedures for their use. (FM 3-0). Offensive IO are capable of degrading an adversary's will to resist and ability to fight. Defensive IO measures passively and actively protect friendly information and C2 systems and limit their vulnerability.

A-6. Offensive and defensive IO are conducted as a fully coordinated effort to ensure the complementary, asymmetric, and reinforcing effects to attack enemy forces, influence others, and protect friendly forces. Relevant information assures that the right person has the right information at the right time for decisionmaking and execution.

THE ELEMENTS OF IO

A-7. Full-spectrum IO incorporates, integrates, and synchronizes traditionally independent capabilities and activities in support of the commander's mission. (See FM 3-0 and FM 3-13.) The Army's doctrinal view of full-spectrum IO as evolving core capabilities are—
- PSYOP.
- OPSEC.
- EW.
- Military deception.
- CNO.
  - CNA.
  - CND.
  - CNE.

A-8. Supporting IO capabilities are—
- Physical destruction.
- Physical security.
- Information assurance.
- Counterpropaganda.
- Counterdeception.
- Counterintelligence.

IO-RELATED ELEMENTS

A-9. The IO-related activities of PA and CMO are closely associated and integrated with the elements of IO as key contributors to information superiority.

INTELLIGENCE SUPPORT TO IO

A-10. All-source intelligence support, encompassing all of its forms, is the principal enabler for successful IO. The commander's capability to synchronize military operations can be heavily influenced by the ability to identify the threat and understand the adversary's capabilities and intentions in the information environment.

A-11. It is essential that the elements of the command's IO capabilities and vulnerabilities are integrated into the command intelligence plan. This leads to COAs that synchronize the elements of IO and IO-related activities into the commander's warfighting plans. Integration of the full-spectrum aspects of IO ensures that the relative importance of information is recognized in the development of a synchronized OPLAN. Further, integration of IO into the planning process provides a methodology for analyzing the threat from a knowledge base that enhances protection of friendly systems and assets while exposing windows of opportunity for attack or exploitation. IPB is the best process we have for understanding the battlefield and the options it presents to friendly and threat forces.

A-12. Provide Intelligence support to IO includes, but is not limited to, the following:

- **Provide Intelligence Support to Offensive IO.** The Intelligence BOS supports offensive IO by providing information to identify critical enemy C2 nodes. Intelligence also helps to identify enemy systems and procedures that may be vulnerable for offensive IO. Additionally, intelligence plays a key role in evaluating and assessing the effectiveness of offensive IO.
  - *Provide Intelligence Support to PSYOP.* This task identifies the cultural, social, economic, and political environment of the AO. It identifies target groups and subgroups and their location, conditions, vulnerabilities, susceptibilities, cultures, attitudes, and behaviors. PSYOP influence foreign target audiences in the AO to support achieving the commander's goals in the AO.
    - Identify the cultural, social, economic, and political environment of the AOI; for example, adversary mechanisms for political control, adversary communication and broadcast systems used to elicit support from the populace, and current and past adversary propaganda activities and their effectiveness.
    - Identify target groups and subgroups and their location, conditions, vulnerabilities, susceptibilities, cultures, attitudes, and behaviors. This includes determining the audience demographics of popular radio and television programs and periodicals; groups influenced by media personalities and political cartoons.
    - Identify impact of planned PSYOP on individuals outside the targeted group (for example, multinational partners, neighboring populations).
  - *Provide Intelligence Support to Military Deception.* This task identifies the capabilities and limitations of the adversary's

intelligence-gathering systems and identifies adversary biases and perceptions.
- Profiles of key adversary leaders.
- Cultural, religious, social, and political characteristics of the country and region.
- Sources of military, economic, or political support.
- Adversary decisionmaking processes, patterns, and biases.
- Adversary perceptions of the military situation in the AO.
- Capabilities and limitations of adversary CI and security services.
- *Provide Intelligence Support to Electronic Attack.* This task supports electronic attack employing jamming, electromagnetic energy, or directed energy against personnel, facilities, or equipment. It identifies critical adversary information systems and C2 nodes. It includes determining and presenting the adversary's electronic OB and their information system infrastructure. The enemy's C2 system vulnerabilities and the means they use to protect their C2 systems is part of the electronic OB.
- **Provide Intelligence Support to Defensive IO.** The Intelligence BOS supports defensive IO by providing information to identify threat IO capabilities and tactics. Intelligence provides information relating to CND, physical security, OPSEC, counterdeception, and counterpropaganda. The Intelligence BOS supports defensive IO by providing information to identify threat IO capabilities and tactics. Intelligence provides information relating to CND. Provide Intelligence Support to OPSEC—identify capabilities and limitations of the adversary's intelligence system to include adversary intelligence objectives and the means, methods, and facilities used by the threat to collect, process, and analyze information—supports the identification of indicators that could be interpreted or pieced together to penetrate EEFI in time to be useful to adversaries.
- **Provide Intelligence Support to Activities Related to IO.** The Intelligence BOS when operating outside US territories supports activities related to IO under some circumstances.
    - *Provide Intelligence Support to CMO.* This task allows military intelligence organizations to collect and provide information and intelligence products concerning foreign cultural, social, economic, and political elements within an IO in support of CMO. Identify cultural, social, economic, and political environment of the AOI, including—
        - Population demographics.
        - Civilian populace attitudes, alliances, and behavior.
        - Availability of basic necessities (food, clothing, water, shelter, medical care) and the ability of the populace to care for itself.
        - Access to medical care.
        - Locations and potential routes, destinations, and assembly areas or sites of displaced persons.
        - Local government type, status, organization, and capabilities.

- Availability of local material and personnel to support military operations.
- NGOs in the AOI, their agenda, resources, and capabilities.

• *Provide Intelligence Support to Public Affairs.* This task identifies the multinational and foreign public physical and social environment, as well as world, HN national, and HN local public opinion, in addition to the propaganda and misinformation capabilities, activities, targets, themes, and dissemination means of the adversary. Identify world, national, and local public opinion (location, biases or predispositions, and agenda of national and international media representatives in the AOI, and trends reflected by the national and international media).

Appendix B
Linguist Support

ROLE OF LINGUISTS

B-1. Military operations are highly dependent on foreign language support. The requirement to communicate with and serve on multinational staffs, communicate with local populations, and exploit enemy forces necessitates the use of linguists. The growing focus on multinational operations increases the competition for limited linguist resources that are vital for mission success. This appendix establishes the framework and process to access, prioritize, and employ the Army's limited organic linguist resources.

LINGUISTIC SUPPORT CATEGORIES

B-2. Foreign language support requirements of US Armed Forces typically fall into one of four broad categories:
- Intelligence and Information Gathering. This category includes the traditional SIGINT and HUMINT disciplines, as well as foreign language support to FP and exploitation of open-source information.
- CMO. This category encompasses all functions relating to military interaction with the civilian population. Foreign language support is critical to CMO in areas such as government liaison, legal agreements, medical support and operations, law enforcement, engineering projects, public safety, security and population control, CA, and PSYOP.
- Logistics. This category consists of foreign language support to sustainment or transportation functions. These include logistical contracting, port, railhead, airhead, or transshipment operations and convoy operations.
- Multinational Operations and Liaison. This category includes the coordination of military operations and liaison with multinational partners, previously unaffiliated nations, and at times adversary or former adversary nations. Multinational operations are becoming more common and increasingly important.

DETERMINING LINGUIST REQUIREMENTS

B-3. To identify linguist requirements, the staff conducts mission analysis and identifies specified or implied tasks requiring foreign language support. Other critical factors are the organization or echelon of command and the location of the mission. The staff uses these criteria to determine the allocation of linguists, such as one linguist team per echelon of command, one linguist per piece of equipment, or one linguist team per location where the function is to be performed. The staff then applies task organization and scheme of maneuver to determine the number of linguists needed for an operation.

B-4. The staff must analyze each linguist assignment to determine the minimum level of foreign language proficiency needed. While interpretation for a peace negotiation requires not only outstanding linguistic capability but also cultural acumen, the translation of routine documents (with the aid of a dictionary) requires a much different skill set. Poor identification of linguist proficiency requirements can tie up the best linguists in less effective roles, creating linguist shortfalls in other areas.

B-5. The relative importance of each of the four linguist support categories is mission dependent. For example, during a NEO civil and military coordination would probably not be as critical as intelligence and information gathering. However, the situation is reversed for a humanitarian assistance mission in which CMOs have a significant impact on mission success. Identifying these "dynamics" helps the commander and staff prioritize linguist requirements.

B-6. Determining linguist requirements for any operation can be difficult because each operation is unique. However, commanders and staffs with a basic knowledge of organic Army linguistic assets, foreign language resource alternatives, and MI skills can successfully assess, prioritize, and employ linguists in support of their military operations.

PLANNING AND MANAGING LINGUIST SUPPORT

B-7. Commanders must consider the linguist requirements as part of their MDMP for every CONPLAN and OPLAN assigned to their commands. Prior staff planning and identification of linguist requirements should prompt commanders to initiate linguist support requests and identify command relationships prior to actual operations. If the mission analysis reveals requirements for linguistic support, the commander must identify what foreign languages are needed, the foreign language proficiency levels needed for each assignment, and the best source of linguists. In addition, if the mission includes intelligence and information collection, the commander must identify MI collection skills required. During mission analysis, the commander should consider linguist requirements for every CONPLAN and OPLAN assigned to his command.

LINGUIST CATEGORIES

B-8. The commander and staff must identify linguist requirements by category:
- Category I – Have native proficiency in the target language (level 4-5) and an advanced working proficiency (Interagency Language Round Table [ILRT] level 2+) in English. They may be locally hired or from a region outside the AO. They do not require a security clearance. They must be screened by the Army CI support team.
- Category II – Are US citizens screened by Army CI personnel and are granted access to SECRET by the designated US government personnel security authority. Have native proficiency in the target language (level 4-5) and an advanced working proficiency (ILRT 2+) in English.
- Category III – Are US citizens screened by Army CI personnel and are granted either TS/SCI clearance or an interim TS/SCI clearance by the

designated US government personnel security authority. Meet a minimum requirement of ILRT level 3. They are capable of understanding the essentials of all speech in a standard dialect. They must be able to follow accurately the essentials of conversation, make and answer phone calls, understand radio broadcasts and news stories, and oral reports (both of a technical and non-technical nature).

PRIMARY STAFF RESPONSIBILITIES

B-9. Primary staff at each echelon has responsibilities for evaluating requirements and managing linguist support. The responsibilities include but are not limited to those discussed below. In addition, each staff section is responsible for determining its linguist support required to meet its operational missions.

Assistant Chief of Staff, G1 (S1):

- Identify linguist requirements needed to support G1/S1 functions in all contingency areas. G1/S1 requirements for linguist support include but are not limited to the following:
  - Coordinate with local authorities on matters of civilian hire, finance, and recordkeeping.
  - Contract for local hire personnel.
  - Coordinate for local morale support and community activities.
  - Coordinate with local authorities for postal operations.
  - Support for administration, counseling, personal affairs, and leave for LN and third-country national (TCN) personnel.
  - Coordinate for local medical support.
  - Liaison with multinational counterparts.
- Linguist staffing and linguist replacement management.
- Identify foreign language skill identifiers for all assigned, attached, or OPCON Army linguists.
- Identify all Army foreign language skilled soldiers not identified on electronic Military Personnel Office System (eMILPO) and Defense Integrated Management Human Resource System (DIMHRS). The Standard Installation Division Personnel System (SIDPERS) was replaced by eMILPO.
- Deploy and provide administrative support of DA and DOD civilian linguists.
- Hire, contract for, and provide administrative support of LN linguists.
- Procure Army foreign language support personnel for screening local labor resources.

Assistant Chief of Staff, G2 (S2):

- Identify linguist requirements needed to support G2/S2 functions in all contingency areas. G2/S2 requirements for linguist support include but are not limited to—
  - Evaluate and/or use local maps and terrain products in operations.
  - Process for MI purposes material taken from EPWs or civilian internees.

- At lower echelons, conduct tactical questioning of refugees, detainees, and EPWs.
- Assess local open-source information for intelligence value.
- Coordinate intelligence and liaison with multinational and HN counterpart.
* Determine, during the initial IPB, all foreign languages (spoken and written) and dialects needed for mission accomplishment.
* Collect, process, produce, and disseminate information derived from linguist sources.
* Provide intelligence training for MI linguists employed in AOs.
* Coordinate for security investigations, as necessary, for local hire linguists.
* Provide support to CI screening of contracted linguists and LN labor force.

Assistant Chief of Staff, G3 (S3):

* Identify linguist requirements needed to support G3/S3 functions in all contingency areas. G3/S3 requirements for linguist support include but are not limited to—
  - Operational coordination and liaison with multinational and HN counterparts.
  - Translate OPORDs and OPLANs for use by multinational counterparts.
* Consolidate unit linguistic requirements and establish priorities.
* Develop linguist deployment and employment plans.
* Develop plans to train linguists and to use linguists for training the force in AO's foreign language survival skills. In addition to global language skills, linguists must have training in specific vocabulary used in the AO; for example, terms used for military, paramilitary, civilian or terrorist organizations, and ethnic groups within the area, nomenclatures of equipment used, and other military or technical vocabulary. Training in the specific dialect used in the AO would also be beneficial.
* Assign, attach, and detach linguists and linguist teams.
* Integrate additional or replacement linguists through operational channels.
* Recommend modernization and development of linguist systems and methods.
* Coordinate mobilization and demobilization of RC linguist support.
* Plan linguist usage for deception operations.
* Plan linguist support for movement of EPWs, detainees, and refugees.
* Coordinate evaluation of linguist support by all staff elements.

Assistant Chief of Staff, G4 (S4):

* Identify linguist requirements needed to support G4/S4 functions in all contingency areas. G4/S4 linguist requirements for linguist support include but are not limited to—

- Procure local supply, maintenance, transportation, and services.
- Coordinate logistics at air and seaports of debarkation.
- Contract with local governments, agencies, and individuals for sites and storage.
- Provide logistical, supply, maintenance, and transportation support to attached linguists.

Assistant Chief of Staff, G5 (S5):

- Identify linguist requirements needed to support G5/S5 functions in all contingency areas. G5/S5 linguist requirements for linguist support include but are not limited to—
  - Determine civilian impact on military operations.
  - Minimize civilian interference with combat operations.
  - Inform civilians of curfews, movement restrictions, and relocations.
  - Provide assistance to liaison with HN and multinational agencies, dignitaries, and authorities.
  - Promote positive community programs to win over support.
  - Determine if multinational operations PSYOP efforts are mutually planned and synchronized.
  - Interpret support to assist resolution of civilian claims against the US Government.
  - Solicit linguistic and cultural knowledge support to protect culturally significant sites.
  - Use linguistic and cultural support to identify cultural and religious customs.
- Assist the G1 in the contracting of local hire linguists.
- Identify foreign language requirements for CMOs.

Assistant Chief of Staff, G6 (S6):

- Identify linguist requirements needed to support G6/S6 functions in all contingency areas. G6/S6 linguist requirements for linguist support include but are not limited to—
  - Coordinate suitable commercial information systems and services.
  - Coordinate with multinational forces on command frequency lists.
  - Coordinate signal support interfaces with HN and multinational forces.
- Manage RF assignments for supporting SIGINT linguist elements.
- Support linguist operations with internal document reproduction, distribution, and message services.
- Integrate automation management systems of linguist units.

SPECIAL STAFF OFFICER RESPONSIBILITIES

B-10. If no special staff officer is assigned the duties below, the corresponding coordinating staff officer should assume those responsibilities. Linguist requirements for special staff officers include but are not limited to the following staff officers.

Liaison Officer:
- Should speak the required foreign language. If not, he requires a translator or interpreter for all aspects of his duties.
- Request interpreters to assist when representing the multinational operations.
- Translate orders, maps, traces, overlays, and documents into multinational foreign languages.

Civilian Personnel Officer:
- Recruit, interview for suitability, and hire civilian local labor force if required.
- Negotiate host country on labor agreements.

Dental Surgeon:
- Administer dental care to support humanitarian mission requirements.
- Rehabilitate, construct, and gain usage of existing dental facilities as required.

Finance Officer:
- Support the procurement process of local goods and services not readily available through normal logistical channels.
- Ensure limited non-US and US pay functions to foreign national, HN, civilian internees, and EPWs are provided.
- Ensure all necessary banking functions are performed in theater.

Surgeon:
- Support medical humanitarian assistance and disaster relief operations.
- Provide medical care of EPWs and civilians within the command's AO.
- Coordinate medical laboratory access in AO.
- Determine the nature of local health threats to the force through populace interviews.
- Determine the identity of local or captured medical supplies.

Veterinary Officer:
- Determine source and suitability of local foods.
- Assist the local population with veterinary service needs.

Chemical Officer:
- Identify enemy force chemical weapons and equipment.
- Communicate NBC risks to supported populations.

Engineer Coordinator:
- Procure proper local materials to support engineering missions.
- Communicate engineering project requirements to contracted local work force.

- Communicate engineering project impact on local landowners and other affected parties.
- Determine, in coordination with G2/S2, suitability of local topographic maps and terrain products.
- Assess environmental concerns of HN and local populations in combined operations.

Provost Marshal:
- Support dislocated and civilian straggler control activities.
- Support internment and resettlement operations, to include displaced civilians.
- Support weapons buy-back programs, as required, and work closely with civil-military liaisons for payments to local officials.
- Support counter-drug and customs activities.
- When authorized, help foreign civil authorities maintain control.
- Conduct liaison with local LEAs.

PSYOP Officer:
- Produce approved PSYOP propaganda and counter-propaganda media.
- Evaluate PSYOP impact on target audience.

Air Defense Coordinator:
- Identify enemy air defense artillery (ADA) weapons and radars.
- Communicate air defense warnings to supported populations.
- Communicate air defense project requirements to contracted local work force.

Safety Officer:
- Provide safety training to local labor force.
- Communicate warnings of dangerous military operations and other hazards to local populace.

Transportation Officer:
- Coordinate commercial and local transportation needs.
- Coordinate movement scheduling and routes with multinational forces and/or HN.

PERSONAL STAFF OFFICER RESPONSIBILITIES

B-11. Personal staff officers are under immediate control of the commander and have direct access to the commander. Most personal staff officers also perform special staff officer duties, working with a coordinating staff officer. These assignments are on a case-by-case basis, depending on the commander's guidance and the nature of the mission; they are very common in stability operations and support operations. Linguist requirements for special staff officers include but are not limited to the following staff officers.

Chaplain:
- Coordinate religious support with multinational partners.
- Determine the impact of local population religious group faiths and practices on military operations.
- Provide religious support to the community to include hospital patients, EPWs, refugees, and civilian detainees.
- Conduct liaison with local population religious leaders in close coordination with the G5.

Public Affairs Officer:
- Act as the commander's spokesman for all communication with external media.
- Assess the accuracy of foreign media interpretation of Public Affairs Office (PAO) releases.
- Assess and recommend news, entertainment, and other information (assisting G5) for contracted services foreign nationals.

Staff Judge Advocate:
- Translate and interpret foreign legal codes, SOFAs, and international laws.
- Determine local environmental laws and treaties through translation services.
- Assess the treatment of EPWs and civilian internees.
- Translate documents to support G4 in local contracts.

SOURCES OF LINGUISTS

B-12. There are various sources that a commander can use to obtain the linguists necessary to support operations. It is vital to know the advantages and disadvantages of each type of linguist and to carefully match the available linguists to the various aspects of the operation.

ARMY LANGUAGE-QUALIFIED MOSs

B-13. The AC MI language-dependent military occupational specialities (MOSs) are 98G with a skill qualification identifier (SQI) of L/352G (Cryptologic Communications Interceptor/ Locator), and their related WO fields. Some soldiers in MOS 96B (All-Source Intelligence Analyst), MOS 97B (CI Agent), MOS 97E/351E (HUMINT Collector) and MOS 98C (SIGINT Analyst), and their related WO fields are trained in foreign languages. Using soldiers in the MOSs mentioned above has many advantages. They are already trained in the military system, are not subject to deployment restrictions (a limiting factor with civilian linguists), have a security clearance and, as US personnel, support the command's interests. The major disadvantage to utilizing these individuals for general foreign language support is that in doing so, they are removed from their primary MI functions. They should be used only in linguistic duties that include intelligence potential. For example, a HUMINT collector (97E) provides linguist support to a medical assistance team as a method to provide access to the local population to determine their attitudes toward US Forces.

B-14. Non-MI Army language qualified MOSs include some enlisted and WOs in career management fields 18 (Special Forces), 37 (PSYOP), 180A (Special Forces); and commissioned officers with a branch code 18 (Special Forces); and functional areas 39 (PSYOP and CA) and 48 (Foreign Area Officer). Particular attention must be paid to the recorded language proficiency and test date of these individuals since the standards vary by field. The same advantages and disadvantages apply as with the AC MI linguists.

B-15. RC language-dependent MOSs include those listed above in the AC. RC linguists have the same set of advantages and disadvantages as listed above for AC language-dependent MOSs. The RC also includes linguists in MOS 97L (translator/interpreter). The 97Ls are specifically trained to be a translator and interpreter. They have the same advantages as the AC linguists. An added advantage is that since their sole job is translation and interpretation, they do not have to be removed from another job in order to be used as a linguist. Their major disadvantage is that they have no additional skill that gives them dual functionality.

Army Linguists Not DOD Trained

B-16. The Army also includes numerous soldiers of all grades who are proficient in a foreign language and are receiving Foreign Language Proficiency Pay (FLPP) but whose primary duties do not require foreign language proficiency. They may have attended a civilian school to learn a foreign language, or they may have acquired proficiency through their heritage. They have the advantage of being trained soldiers and are therefore readily deployable to all areas of the battlefield.

B-17. These soldiers may have the specific vocabulary and military skill knowledge for certain linguist support missions. For example, a supply sergeant who speaks the local language would be an invaluable asset to the G4. There are disadvantages in that they already have another job and units are reluctant to give up personnel especially if they are in key positions. Their capabilities are difficult to assess. Since they are not required to take the Defense Language Proficiency Test (DLPT) if they are not receiving FLPP, it is often difficult for the G1/S1 to identify them as a linguist or for a non-linguist to judge the level of their foreign language capability.

Other Service Linguists

B-18. Other service linguists have the advantage of deployability, loyalty, and clearance, but must often learn the Army system and specific Army vocabulary. They are also difficult to obtain since their parent service probably also lacks a sufficient number of trained linguists. Other service linguists, however, will be valuable in joint operation centers and joint activities. When serving the JTF headquarters, Army commanders and staffs must be aware of the linguists in the other services in order to plan for the participation and optimize their employment.

US Contract Linguists

B-19. US civilians can be contracted to provide linguist support. They have an advantage over LN hires in that their loyalty to the US is more readily evaluated, and it is easier for them to be granted the necessary security clearance. However, there are usually severe limitations on the deployment and use of civilians. A careful assessment of their language ability is important because, in many cases, they use "old fashioned" terms, or interject US idioms. If the linguists are recent émigrés, the use of the language in their country of origin could be dangerous to them, or their loyalty may reside with their own country when at odds with US interests.

Multinational Linguists

B-20. Multinational linguists have their own set of advantages and disadvantages. These linguists may be unfamiliar with the US military system unless they have previously participated in a multinational operation with US forces. They may have a security clearance, but clearances are not necessarily equal or reciprocal, automatically guaranteeing access to classified or sensitive information between nations. They support the command's interest but may have differing priorities or responsibilities within their assigned AOs. These linguists also are already fulfilling specific duties for their own nation, which may also have a shortage of linguists. The major disadvantage to acquiring and maintaining multinational linguist support is that they are outside the C2 (via military authority or military contract) of the US forces. These linguists will be valuable in multinational operations centers and activities.

Local National Contract Linguists

B-21. LN hires will provide the bulk of your linguist support. They are usually less expensive to hire than US civilians and will know the local dialect, idioms, and culture. The expertise of these linguists in particular areas or subject matters can be an asset. However, there are several potential problems with using LN hires, to include limited English skills and loyalty considerations. Therefore, a screening interview or test is necessary to determine their proficiency in English. These individuals must also be carefully selected and screened by CI personnel (with US linguist support) initially and periodically throughout their employment. Their loyalty is always questionable. Local prejudices may influence them, and they may place their own interests above those of the US.

EVALUATING LINGUIST PROFICIENCY

B-22. Commanders and staffs must understand the Army linguist proficiency evaluation system in order to effectively plan for and employ linguists. Evaluation and reevaluation of linguist proficiency is covered in detail in AR 611-6, Section III. Language testing is required for all Army personnel in a language-dependent MOS, who have received foreign language training at government expense, who are receiving FLPP, or who are in a language-required position regardless of MOS. Other Army personnel who have knowledge of a foreign language are encouraged to take the proficiency test and may work as linguists.

B-23. The Army uses the DLPT to determine foreign language proficiency levels. DLPTs are listed by foreign language in DA Pam 611-16. In foreign languages where no printed or recorded test exists, oral interview tests are arranged. The DLPT is an indication of foreign language capability, but it is not the definitive evaluation of an individual's ability to perform linguist support.

B-24. AR 611-6, Appendix D, Sections 1 through 4, describes the proficiency levels for the skills of speaking, listening, reading, and writing a foreign language based on the interagency roundtable descriptions. The plus-level designators, shown as a "+" symbol, are used to designate when a linguist is above a base level, but not yet to the capability of the next level. For example, 2+ would indicate a better than limited working proficiency in the foreign language. The six "base levels" of proficiency, as established by DLPT and/or oral exam, are—

- Level 0 (No proficiency). The soldier has no functional foreign language ability. Level 0+. The minimum standard for Special Forces personnel indicates a memorized proficiency only.
- Level 1 (Elementary proficiency). The soldier has limited control of the foreign language skill area to meet limited practical needs and elementary foreign language requirements.
- Level 2 (Limited working proficiency). The linguist is sufficiently skilled to be able to satisfy routine foreign language demands and limited work requirements.
- Level 3 (General professional proficiency). The linguist is capable of performing most general, technical, formal, and informal foreign language tasks on a practical, social, and professional level.
- Level 4 (Advanced professional proficiency). The linguist is capable of performing advanced professional foreign language tasks fluently and accurately on all levels.
- Level 5 (Functionally native proficiency). The linguist is functionally equivalent to an articulate and well-educated native in all foreign language skills; and reflects the cultural standards of the country where the foreign language is natively spoken.

B-25. The above proficiency base levels designate proficiency in any of the four language skills: listening, reading, speaking, and writing. The most evaluated skills on the DLPT are reading and listening. These tests are not for use in evaluating linguists above the 3 proficiency level. Most Army linguist DLPT scores show only two skill levels: listening and reading (for example, 2+/3, or 3/1+).

SUSTAINING MILITARY LINGUIST PROFICIENCY

B-26. Language proficiency diminishes with lack of use and absence of exposure to the foreign language. To ensure combat readiness, commanders should require all military linguists receive periodic language training. In-country language immersion training, in-garrison contracted language instructors, on-line foreign newspapers, and foreign radio broadcasts are all examples of language training resources. Funding for language training is available through MACOM language training program funds.

Glossary

The glossary lists acronyms and terms with Army or joint definitions, and other selected terms. Where Army and joint definitions are different, (Army) follows the term. Terms for which FM 2-0 is the proponent FM (authority) are marked with an asterisk (*) and followed by the number of the paragraph (¶) where they are discussed. For other terms, the number of the proponent FM follows the definition. JP 1-02 and FM 1-02 are posted in the Joint Electronic Library, which is available online and on CD-ROM.

- Use this URL to access JP 1-02 online: http://www.dtic.mil/doctrine/jel/doddict/.
- Use this URL to access FM 1-02 online: http://www.dtic.mil/doctrine/jel/service_pubs/101_5_1.pdf.
- Follow this path to access JP 1-02 on the Joint Electronic Library CD-ROM: Mainmenu>Joint Electronic Library>DOD Dictionary.
- Follow this path to access FM 1-02 on the Joint Electronic Library CD-ROM: Mainmenu>Joint Electronic Library>Service Publications>Multiservice Pubs> FM 101-5-1.

A2C2	Army Airspace Command and Control
AA	avenue of approach
AAMDC	Area Air and Missile Defense Command
ABCS	Army Battle Command System
AC	Active Component
ACE	analysis and control element
ACINT	acoustic intelligence (water)
ACOUSTINT	acoustic intelligence (atmosphere)
ACR	armored cavalry regiment
ACT	analysis and control team
ADA	air defense artillery
ADCON	administrative control
ADDO	assistant deputy director for operations
ADO	air defense officer
ADP	automated data processing
AFACSI	Air Force Assistant Chief of Staff for Intelligence
AFIWC	Air Force Information Warfare Center
AFMIC	Armed Forces Medical Intelligence Center

AFOSI	Air Force Office of Special Investigation
AGM	attack guidance matrix
AIA	Air Force Intelligence Agency
AITR	aided target recognition
ALOC	administrative and logistics operations center
AMC	US Army Materiel Command
AMD	Air Missile Defense
AMHS	Automated Message Handling System
AMIB	Allied Military Intelligence Battalion
*analysis	Determination of the significance of the information, relative to information and intelligence already known, and drawing deductions about the probable meaning of the evaluated information. (¶1-58)
AO	area of operations
AOC	area of concentration
AOI	area of interest
AOIR	area of intelligence responsibility
AOR	area of responsibility
AR	Army Regulation
ARAT	Army Reprogramming Analysis Team
ARCENT	US Army Central Command
ARFOR	Army forces
ARISC	Army Reserve Intelligence Support Center
ARL	airborne reconnaissance low
ARNG	US Army National Guard
ARSOF	Army Special Operations Forces
ASCC	Army Service Component Commander
ASPO	Army Space Program Office
AT	antiterrorism
ATCAE	Army Technical Control Analysis Element
ATCCS	Army Tactical Command and Control System
ATR	automatic target recognition
ART	Army tactical task
BDA	battle damage assessment
BDU	battle-dress uniform

BFACS	Battlefield Functional Area Control System
C/J2	combined J2
C2	command and control
CA	civil affairs
CASO	Collection Activities and Source Operations
CAT	crisis action team
CBRN	chemical, biological, radiological, and nuclear
CBRNE	chemical, biological, radiological, nuclear, and high-yield explosive
CCD	coherent change detection
CCIR	commander's critical information requirement
CD-ROM	compact disc-read only memory
CE	counterespionage
CECOM	Communications-Electronics Command
CED	captured enemy document
CEE	captured enemy equipment
CEM	captured enemy materiel
CFSO	counterintelligence force protection source operations
CGS	common ground station
CI	counterintelligence
CIA	Central Intelligence Agency
CIAC	counterintelligence analysis cell
CICA	Counterintelligence Coordinating Authority
CITO	Central Imagery Tasking Office
CJCS	Chairman, Joint Chiefs of Staff
CMEC	Captured Materiel Exploitation Center
CMF	career management field
CMISE	Corps Military Intelligence Support Element
CMO	civil-military operations
CNA	computer network attack
CND	computer network defense
CNE	computer network exploitation
CNN	Cable News Network
CNO	computer network operations

FM 2-0, C1

COA	course of action
COCOM	combatant command (command authority)
COG	center of gravity
COMINT	communications intelligence
COMSEC	communications security
CONOPS	contingency operations
CONPLAN	contingency plan
CONUS	continental United States
COP	common operational picture
COSCOM	Corps Support Command
*counterintelligence	1. Information gathered and activities conducted to protect against espionage, other intelligence activities, sabotage, or assassinations conducted by or on behalf of foreign governments or elements thereof, foreign organizations, or foreign persons, or international terrorist activities. (JP 1-02) 2. (Army) Counterintelligence counters or neutralizes intelligence collection efforts through collection, CI investigations, operations, analysis and production, and functional and technical services. CI includes all actions taken to detect, identify, exploit, and neutralize the multidiscipline intelligence activities of friends, competitors, opponents, adversaries, and enemies; and is the key intelligence community contributor to protect US interests and equities. (FM 2-0) (¶1-111)
CP	command post
CRN	command radio net
CS	combat support
CSG	cryptologic support group
CSS	combat service support
CSSCS	Combat Service Support Control System
CT	Counterterrorism
*cueing	The use of one or more sensor systems to provide data that directs collection by other systems. (¶2-53)
CWT	combat weather team
DA	Department of the Army
DAC	Department of the Army Civilian
DAPDM	Department of the Army Production and Dissemination Management
DCI	Director of Central Intelligence
DCS, G2	Deputy Chief of Staff, Intelligence

FM 2-0, C1

*debriefing	The systematic questioning of individuals to procure information to answer specific collection requirements by direct and indirect questioning techniques. (¶6-4)
DF	direction finding
DHS	Defense HUMINT Service
DIA	Defense Intelligence Agency
DIAC	Defense Intelligence Analysis Center
DIMHRS	Defense Integrated Management Human Resource System
DIO	Defensive Information Operations
DISCOM	Division Support Command
DISUM	daily intelligence summary
DLPT	Defense Language Proficiency Test
DMS	Director of Military Support
DOCEX	document exploitation
*document exploitation	The systematic extraction of information from all media formats in response to collection requirements. (¶6-4)
DOD	Department of Defense
DOJ	Department of Justice
DOT	Department of Transportation
DP	decision point
DPM	dissemination program manager
DS	direct support
DSM	decision support matrix
DST	decision support template
EA	electronic attack
EAC	echelons above corps
ECOA	enemy course of action
EEFI	essential elements of friendly information
ELINT	electronic intelligence
EM	electromagnetic
eMILPO	electronic Military Personnel Office System
EMS	electromagnetic spectrum
E-O	electro-optical
EOB	electronic order of battle
EOD	explosive ordinance disposal

EP	electronic protection
EPW	enemy prisoner of war
ES	electronic warfare support (Army)
*evaluating	1. Comparing relevant information (RI) on the situation or operation against criteria to determine success or progress. (FM 6-0) 2. In intelligence usage, appraisal of an item of information in terms of credibility, reliability, pertinence, and accuracy. (FM 2-0) (¶1-26 and 1-98)
EW	electronic warfare
EWS	electronic warfare support (joint)
FBI	Federal Bureau of Investigation
FDA	functional damage assessment
FDU	Force Design Update
FEMA	Federal Emergency Management Agency
FFIR	friendly force information requirement
FISINT	foreign instrumentation signals intelligence
FISS	Foreign Intelligence Security Service
FLPP	Foreign Language Proficiency Pay
FM	Field Manual
FMA	foreign materiel acquisition
FME	foreign materiel exploitation
FMEP	Foreign Materiel Exploitation Program
FMP	Foreign Materiel Program
FMT	Foreign Materiel for Training
FP	force protection
FRAGO	fragmentary order
FS	fire support
FSAC	Fire Support Armaments Center
FSE	fire support element
FSO	fire support officer
G1	Assistant Chief of Staff, Personnel
G2	Assistant Chief of Staff, Intelligence
G3	Assistant Chief of Staff, Operations
G4	Assistant Chief of Staff, Logistics
G5	Assistant Chief of Staff, Civil/Military Affairs

G6	Assistant Chief of Staff, Command, Control, Communications, and Computer Operations
G7	Assistant Chief of Staff, Information Operations
GEOINT	geospatial intelligence
GMI	general military intelligence
GPS	Global Positioning System
GS	general support
GS-R	general support-reinforcing
HN	host nation
HOC	HUMINT operations cell
HPT	high-payoff target
HQ	headquarters
HQDA	Headquarters, Department of the Army
HSI	hyperspectral imaging
HUMINT	human intelligence
*human intelligence	Collection by a trained HUMINT collector of foreign information from people and multimedia to identify elements, intentions, composition, strength, dispositions, tactics, equipment, personnel, and capabilities. It uses human sources and a variety of collection methods, both passively and actively, to gather information to satisfy the commander's intelligence requirements and cross-cue other intelligence disciplines. (¶1-106 and ¶6-1)
HVT	high-value target
I/R	internment and resettlement (operations)
I&W	indications and warnings
ICC	Intelligence Coordination Center (US Coast Guard)
ICF	intelligence contingency fund
ICL	intelligence coordination line
IDC	Information Dominance Center
IIR	intelligence information report
IM	information management
IMA	individual mobilization augmentee
IMEP	International Materiel Evaluation Program
IMETS	Integrated Meteorological System
IMINT	imagery intelligence

*indications and warning	(joint) Those intelligence activities intended to detect and report time-sensitive intelligence information on foreign developments that could involve a threat to the United States or allied and/or coalition military, political, or economic interests or to US citizens abroad. It includes forewarning of enemy actions or intentions; the imminence of hostilities; insurgency; nuclear or non-nuclear attack on the United States, its overseas forces, or allied and/or coalition nations; hostile reactions to US reconnaissance activities; terrorists attacks; and other similar events. (¶2-19)
*indicator	(Army) Positive or negative evidence of threat activity or any characteristic of the AO which points toward threat vulnerabilities or the adoption or rejection by the threat of a particular capability, or which may influence the commander's selection of a COA. Indicators may result from previous actions or from threat failure to take action. (FM 2-0) (¶5-14)
INFOSEC	information security
*infrastructure	1. (joint) All building and permanent installations necessary for support, redeployment, and military forces operations (e.g., barracks, headquarters, airfields, communications facilities, stores, port installations, and maintenance stations). (JP 4-01.8) 2. (Army) In intelligence usage, the basic underlying framework or feature of a thing; in economics, basic resources, communications, industries, and so forth, upon which others depend; in insurgency, the organization (usually hidden) of insurgent leadership. (FM 2-0) (¶5-8)
INR	Bureau of Intelligence and Research, Department of State
INS	Immigration and Naturalization Service
INSCOM	US Army Intelligence and Security Command
intelligence	1. (joint) The product resulting from the collection, processing, integration, analysis, evaluation, and interpretation of available information concerning foreign countries or areas. 2. Information and knowledge about an adversary obtained through observation, investigation, analysis, or understanding. (¶1-3)
*intelligence coordination line	A line that designates the boundary between AOIRs. The G2/S2 establishes ICLs to facilitate coordination between higher, lateral, and subordinate units; coordinates with the G3/S3 to direct subordinates to track enemy units and HPTs in their areas; and hands over intelligence responsibility for areas of the battlefield. The establishment of ICLs ensures that there are no gaps in the collection effort; that all echelons are aware of the location, mission, and capabilities of other assets; facilitates asset cueing, and provides timely exchange of information between assets. The G2/S2 keeps abreast of collection activities in progress (all echelons) and battlefield developments through the ICLs. (¶3-13)

intelligence discipline	(joint) A well-defined area of intelligence collection, processing, exploitation, and reporting using a specific category of technical or human resources. There are seven major disciplines: human intelligence, imagery intelligence, measurement and signature intelligence, signals intelligence (communications intelligence, electronic intelligence, and foreign instrumentation signals intelligence), all-source intelligence, technical intelligence, and counterintelligence. (¶1-104)
*intelligence process	A theoretical model used to describe intelligence operations. The intelligence process is not a framework for actual operations because the functions and common tasks occur in parallel as opposed to sequentially. (¶1-91)
*intelligence products	Intelligence products are generally placed in one of six categories: I&W, current, general military, target, S&T, and CI. The categories are distinguished from each other primarily by the purpose for which the intelligence was produced. The categories often overlap, and the same intelligence data can be used in each of the categories. (¶1-5)
INTREP	intelligence report
INTSUM	intelligence summary
*intelligence requirements	(Army) Those requirements generated from the staff's IRs regarding the enemy and environment that are not a part of the CCIR (PIR and FFIR). Intelligence requirements require collection and can pro-vide answers in order to identify indicators of enemy actions or intent, which reduce the uncertainties associated with an operation. Significant changes (i.e., branches and sequels) with an operation usually lead to changes in intelligence requirements. (¶2-7)
*intelligence running estimate	The intelligence running estimate is a continuous flow and presentation of relevant information and predictive intelligence that, when combined with the other staff running estimates, enable the decisionmaker's visualization and situational understanding of the AOI in order to achieve information superiority. The intelligence running estimate requires constant verification to support situational understanding of the current situation as well as predictive assessments for future operations. (¶5-30)
*intelligence synchronization	The task that ensures ISR operations are linked to the commander's requirements and respond in time to influence decisions and operations. The intelligence officer, with staff participation, synchronizes the entire collection effort to include all assets the commander controls, assets of lateral units and higher echelons units and organizations, and intelligence reach to answer the commander's CCIR (PIR and FFIR). (¶1-24)

FM 2-0, C1

*intelligence synchronization plan	The plan the intelligence officer uses, with staff input, to synchronize the entire collection effort to include all assets the commander controls, assets of lateral units, and higher echelon units and organizations, and intelligence reach to answer the commander's CCIR (PIR and FFIR). (¶1-38)
intelligence warfighting function	The related tasks and systems that facilitate understanding of the operational environment
IO	information operations
IPB	intelligence preparation of the battlefield
IPIR	Initial Phase Imagery Report
IPTF	international police task force
IR	information requirement
IRINT	infrared intelligence
IRR	Individual Ready Reserve
ISM	intelligence synchronization matrix
ISR	intelligence, surveillance, and reconnaissance
J2	Joint Staff Directorate, Intelligence
J2A	Joint Staff Directorate, Intelligence (Administrative)
J2J	Joint Staff Directorate, Intelligence (Joint)
J2M	Joint Staff Directorate, Intelligence (Management)
J2O	Joint Staff Directorate, Intelligence (Operations)
J2P	Joint Staff Directorate, Intelligence (Assessments, Doctrine)
J2T	Joint Staff Directorate, Intelligence (Targeting)
J2X	Joint Staff Directorate, Intelligence (CI & HUMINT)
J3	Joint Staff Directorate, Operations
J5	Joint Staff Directorate, Civil Affairs
JAC	Joint Analysis Center
JCMEC	Joint Captured Materiel Exploitation Center
JCO	Joint Commission Observers
JCS	Joint Chiefs of Staff
JDEC	Joint Document Exploitation Center
JDISS	Joint Deployable Intelligence Support System
JFC	joint force commander
JIAWG	joint interagency working group
JIC	Joint Intelligence Center

JIF	Joint Interrogation Facility
JISE	joint intelligence support element
JOA	joint operations area
JP	Joint Publication
JSTARS	Joint Surveillance Target Attack Radar System
JTF	joint task force
JTMD	joint table of mobilization and distribution
JWICS	Joint Worldwide Intelligence Communications System
LAN	local area network
LASINT	laser intelligence
LCC	land component commander
LEA	law enforcement agency
LIDAR	laser radar
LN	local national
LNO	liaison officer
LOC	line of communication
LOS	line of sight
LPT	logistics preparation of the theater
LTIOV	latest time information is of value
MACOM	major Army command
MASINT	measurement and signature intelligence
*materiel	(Army) In intelligence usage, the all-encompassing term for the weapons systems, equipment, apparatus, documents, and supplies of a foreign military force or nonmilitary organization. (¶2-24)
MCIA	Marine Corps Intelligence Agency
MCOO	modified combined obstacle overlay
MCS	Maneuver Control System
MDMP	military decisionmaking process
MEA	munitions effects assessment
METL	mission-essential task list
METT-TC	mission, enemy, terrain and weather, troops and support available, time available, civil considerations
MI	military intelligence
MLT	machine language translation

MNFC	multinational force commander
MO	method of operation
MOPP	mission-oriented protective posture
MOS	military occupational specialty
MP	military police
MSC	major subordinate command
MSE	mobile subscriber equipment
MSI	multi-spectral imagery
MSIC	Missile and Space Intelligence Center
MTI	moving target indicator
MTOE	modified table of organization and equipment
MTW	major theater war
NAI	named area of interest
NAIC	National Air Intelligence Center
NATO	North Atlantic Treaty Organization
NBC	nuclear, biological, and chemical
NCA	National Command Authorities
NCO	noncommissioned officer
NCR	national cryptologic representative
NCTI	non-cooperative target identification
NEO	noncombatant evacuation operation
NFH	National Defense Headquarters, Canada
NGA	National Geospatial-Intelligence Agency
NGO	non-governmental organization
NIC	National Intelligence Center
NIIRS	National Imagery Interpretability Rating Scale
NIMA	National Imagery and Mapping Agency
NIST	National Intelligence Support Team
NLT	not later than
NMCC	National Military Command
NMIC	National Maritime Intelligence Center
NMJIC	National Military Joint Intelligence Center
NRO	National Reconnaissance Office
NRT	near-real time

NSA	National Security Agency
NSC	National Security Council
NUCINT	Nuclear Intelligence
NVD	night vision device
OB	order of battle
OE	operational environment
OMA	Office of Military Affairs
OMB	Office of Management and Budget
OMT	operational management team
ONDCP	Office of National Drug Control Policy
ONI	Office of Naval Intelligence
ONIR	overhead non-imaging infrared
OPCON	operational control
OPLAN	operation plan
OPORD	operation order
OPSEC	operations security
OSC	operations support cell
OSCE	Organization for Security and Cooperation in Europe
OSD	Office of the Secretary of Defense
OSINT	open-source intelligence
PAO	Public Affairs Office
PDA	physical damage assessment
PIAP	Police Information Assessment Program
PIO	police intelligence operations
PIR	priority intelligence requirement
PME	peacetime military engagement
POC	point of contact
POD	port of debarkation
POE	port of embarkation
*predictive intelligence	Intelligence analysis conclusions, assessments or products that attempt to define, describe, present, or portray the future situation or conditions of the enemy, +terrain, weather and civil considerations. (¶1-3)
*priority intelligence requirements	(joint) Those intelligence requirements for which a commander has an anticipated and stated priority in his task of planning and decision-making. (FM 2-0) (¶1-32)

*process/ processing	A function of the intelligence process that involves converting collected data, which is not already in a comprehensible form when it is reported, into a form that is understandable and suitable for analysis and production of intelligence. Examples of processing include developing film, enhancing imagery, translating a document from a foreign language, converting electronic data into a standardized report that can be analyzed by a system operator, and correlating dissimilar or jumbled information by assembling like elements before the information is forwarded for analysis. (FM 2-0) (¶1-96)
*production	In intelligence usage, conversion of information into intelligence through the integration, analysis, evaluation, and interpretation of all source data and the preparation of intelligence products in support of known or anticipated user requirements. (¶4-20)
PSYOP	psychological operations
QRC	quick reaction capability
QRF	quick reaction force
R&D	research and development
RADINT	radar intelligence
RC	reserve component
RDEC	Research, Development, and Engineering Center
*reconnaissance, surveillance, and target acquisition/intelligence surveillance and reconnaisssance	A full spectrum combined arms mission that integrates ground and air capabilities to provide effective, dynamic, timely, accurate and assured combat information and multi-discipline actionable intelligence for lethal and non-lethal effects/decisions in direct support of the ground tactical commander.
REMBASS	Remotely Monitored Battlefield Sensor System
*requirements management	The intelligence task that develops a prioritized list of what information needs to be collected and produced into intelligence, dynamically updates and adjusts those requirements in response to mission adjustments/changes, and places a latest time intelligence is of value to ensure intelligence and information is reported to meet operational requirements. (¶2-83)
RF	radio frequency
RFI	request for information
RFW	radio frequency weapon
RI	relevant information
RM	requirements management
ROE	rules of engagement
ROK	Republic of Korea
ROZ	restricted operating zone

RSO&I	reception, staging, onward movement, and integration
RSOC	Regional Security Operations Center
RSTA	reconnaissance, surveillance, and target acquisition
RSP	render safe procedure
RTCAE	Regional Technical Control and Analysis Element
running estimate	A staff estimate, continuously updated, based on new information as the operation proceeds. (FM 6-0) (¶5-28)
S&T	scientific and technical
S&TI	scientific and technical intelligence
S1	Staff Officer, Personnel
S2	Staff Officer, Intelligence
S2X	Staff Officer, Intelligence (CI & HUMINT)
S3	Staff Officer, Operations
S4	Staff Officer, Logistics
S5	Staff Officer, Civil Affairs
S6	Staff Officer, Command, Control, Communications, and Computer Operations
SAP	special access program
SAR	synthetic aperture radar
SATCOM	satellite communications
SBCCOM	Chemical and Biological Defense Command
SCI	sensitive compartmented information
SCIF	sensitive compartmented information facility
SIDPERS	Standard Installation Division Personnel System
SIGINT	signals intelligence
SIGO	signal officer
SII	statement of intelligence interest
SIO	senior intelligence officer
SIPRNET	Secret Internet Protocol Router Network
SIR	specific information requirement
SITMAP	situation map

FM 2-0, C1

*situation template	(Army) A depiction of a potential adversary course of action as part of a particular adversary operation. Situation templates are developed on the adversary's current situation (for example, training and experience levels, logistic status, losses, and disposition), the environment, and adversary doctrine or patterns of operations. The commander dictates the level to depict the adversary based on the factors of METT-TC (at minimum two levels of command below the friendly force) as a part of his guidance for mission analysis. (FM 2-0) (¶1-58)
SJA	staff judge advocate
SMU	special mission unit
SOCOM	satellite communications on-the-move
SOF	special operations forces
SOFA	status of forces agreement
SOP	standing operating procedure
SOTA	SIGINT operational tasking authority
SR	special reconnaissance
SRTD	signals research and target development
SSO	special security office
STANAG	standardization agreement (NATO)
STICEUR	Scientific and Technology Intelligence Center Europe
SUPR	supplemental imagery report
SWO	staff weather officer
TA	target acquisition
TACON	tactical control
*tactical questioning	The expedient initial questioning for information of immediate tactical value. (FM 2-0) (¶6-6)
TAI	target area of interest
TCAE	technical control and analysis element
TCN	third-country national
*technical control	The technical function to ensure adherence to existing policies or regulations and provide technical guidance for MI activities, particularly HUMINT, SIGINT, and CI operations. (FM 2-0) (¶3-23)
TECHINT	technical intelligence
TECOM	Test and Evaluation Command
TENCAP	Tactical Exploitation of National Capabilities
TEU	Technical Escort Unit

THREATCON	threat condition
TIB	Theater Intelligence Brigade
TIG	Theater Intelligence Group
TIM	toxic industrial material
TLP	troop-leading procedure
TOC	tactical operations center
TPFDD	Time Phased Force Deployment Data
TRADOC	US Army Training and Doctrine Command
TSA	target system assessment
TSCM	Technical Surveillance Countermeasures
TSE	theater support element
TTP	tactics, techniques, and procedures
TVA	threat vulnerability assessment
UAV	unmanned aerial vehicle
UN	United Nations
US	United States
USAF	United States Air Force
USAR	United States Army Reserve
USC	United States Code
USCENTCOM	United States Central Command
USCG	United States Coast Guard
USCS	United States Cryptologic System
USEUCOM	United States European Command
USJFCOM	United States Joint Forces Command
USMC	United States Marine Corps
USNORTHCOM	United States Northern Command
USPACECOM	United States Space Command
USPACOM	United States Pacific Command
USSID	United States Signal Intelligence Directive
USSOCOM	United States Special Operations Command
USSOUTHCOM	United States Southern Command
USSPACECOM	United States Space Command
USSTRATCOM	United States Strategic Command
USTRANSCOM	United States Transportation Command

VJ2	Vice J2
VTC	video teleconference
WAN	wide area network
WARNO	warning order
WMD	weapons of mass destruction
WO	warrant officer

Bibliography

The bibliography lists field manuals published with new numbers followed by old number.

DOCUMENTS NEEDED

These documents must be available to the intended users of this publication.

JP 1-02. *Department of Defense Dictionary of Military and Associated Terms.* June 1999. [Online] Available http://www.dtic.mil/doctrine/jel/doddict/

FM 1-02/MCRP 5-2A. *Operational Terms and Symbols* (Drag Draft). 21 February 2003.

READINGS RECOMMENDED

These sources contain relevant supplemental information.

ARMY PUBLICATIONS

Most Army doctrinal publications are available online: http://155.217.58.58/atdls.htm

AR 381-10. *US Army Intelligence Activities.* 1 July 1984.

AR 381-12. *Subversion and Espionage Directed Against the U.S. Army (SAEDA).* 15 January 1993.

AR 381-20. *The Army Counterintelligence Program.* 15 November 1993.

AR 611-6. *Army Linguist Management.* 16 February 1996.

AR 715-9. *Contractors Accompanying the Force.* 29 October 1999.

FM 1 (100-1). *The Army.* 14 June 2001.

FM 2-19.402. *STRYKER Brigade Combat Team Intelligence Operations.* 1 March 2003.

FM 2-19.602. *Surveillance Troop.* 1 March 2003.

FM 2-33.5/ST. *Intelligence Reach Operations.* 1 June 2001.

FM 3-0. *Operations.* 14 June 2001.

FM 3-05.102. *Army Special Operations Forces Intelligence.* 31 August 2001 (formerly FM 34-36).

FM 3-05.30. *Psychological Operations.* 19 June 2000.

FM 3-06.11. *Combined Arms Operations in Urban Terrain.* 28 February 2002.

FM 3-13. *Information Operations.* 28 November 2003 (formerly FM 100-6).

FM 3-19. *NBC Reconnaissance.* 19 November 1993.

FM 3-19.1. *Military Police Operations.* 22 March 2001.

FM 3-19.4. *Military Police Leaders' Handbook.* 4 March 2002.

FM 3-19.40. *Military Police Internment/Resettlement Operations.* 30 September 1987.

FM 3-20.98. *Reconnaissance Platoon.* May 2001.

FM 34-2. *Collection Management and Synchronization Planning.* 8 March 1994.

FM 34-52. (FM 2-22-3) *Human Intelligence.* 28 September 1992.

FM 3-90. *Tactics.* 4 July 2001.

FM 3-90.3. *The Mounted Brigade Combat Team.* 1 November 2001.

FM 3-100.4. *Environmental Considerations in Military Operations.* 15 June 2000, Change 1, 11 May 2001.

FM 3-100.21. *Contractors on the Battlefield.* 26 March 2000.

FM 5-170. *Engineer Reconnaissance.* 5 May 1998.

FM 6-0. *Mission Command: Command and Control of Army Forces.* 11 August 2003.

FM 6-20-10. *Tactics, Techniques, and Procedures for the Targeting Process.* 8 May 1996.

FM 6-121. *Tactics, Techniques, and Procedures for Field Artillery Target Acquisition.* 25 September 1990.

FM 7-0. *Training the Force.* 22 October 2002.

FM 7-10. *The Infantry Rifle Company.* 14 December 1990.

FM 7-15. *Army Universal Task List.* 31 August 2003.

FM 7-92. *The Infantry Reconnaissance Platoon and Squad (Airborne, Air Assault, Light Infantry).* 23 December 1992.

FM 7-100. *Opposing Force Doctrinal Framework and Strategy.* 1 May 2003.

FM 17-97. *Cavalry Troop.* 3 October 1995.

FM 17-98. *Scout Platoon.* 10 April 1999.

FM 19-1. *Military Police Support for the Airland Battle.* 23 May 1988.

FM 27-10. *The Law of Land Warfare.* 18 July 1956.

FM 31-20-5. *Special Reconnaissance Tactics, Techniques, and Procedures for Special Forces.* 23 March 1993.

FM 34-2-1. *Tactics, Techniques, and Procedures for Reconnaissance and Surveillance and Intelligence to Counterreconnaissance.* 19 June 1991.

FM 34-3. *Intelligence Analysis.* 15 March 1990.

FM 34-8-2. *Intelligence Officer's Handbook.* 1 May 1998.

FM 34-10. *Division Intelligence and Electronic Warfare Operations.* 25 November 1986.

FM 34-25. *Corps Intelligence and Electronic Warfare Operations.* 30 September 1987.

FM 34-37. *Echelons Above Corps (EAC) Intelligence and Electronic Warfare (IEW) Operations.* 15 January 1991.

FM 34-54. *Technical Intelligence.* 30 January 1998.

FM 34-80. *Brigade and Battalion Intelligence and Electronic Warfare Operations.* 15 April 1986.

FM 34-130. *Intelligence Preparation of the Battlefield.* 8 July 1994.

FM 4-0. *Combat Service Support.* 29 August 2003.

FM 41-10. *Civil Affairs Operations.* 14 February 2000.

FM 100-25. *Doctrine for Army Special Operations Forces.* August 1999.

FM 101-5. *Staff Organization and Operations.* 31 May 1997.

FM 101-5-1. *Operational Terms and Graphics.* 30 September 1997.

FM 101-5-2. *US Army Report and Message Formats.* 29 June 1999.

ST 2-22.7. *Tactical Human Intelligence and Counterintelligence Operations.* April 2002.

ST 2-33.5. *Intelligence Reach Operations.* 1 June 2001.

ST 2-50. *Intelligence and Electronic Warfare Assets.* June 2002.

ST 2-50.4. *Combat Commanders' Handbook on Intelligence.* September 2001 (formerly FM 34-81, 28 September 1992).

JOINT PUBLICATIONS

Most joint publications are available online: http://www.dtic.mil/doctrine/jel/

JP 0-2. *Unified Action Armed Forces (UNAAF).* 10 July 2001.

JP 1. *Joint Warfare for the US Armed Forces.* 14 November 2000.

JP 1-06. *Joint Tactics, Techniques, and Procedures for Financial Management During Joint Operations.* 22 December 1999.

JP 2-0. *Joint Doctrine for Intelligence to Operations.* January 2000.

JP 2-01. *Joint Intelligence Support to Military Operations.* 20 November 1996.

JP 2-01.1. *Joint Tactics, Techniques, and Procedures for Intelligence Support to Targeting.* 9 January 2003.

JP 2-01.3. *Joint Tactics, Techniques, and Procedures for Joint Intelligence Preparation of the Battlespace*, 24 May 2000.

JP 2-02. *National Intelligence to Joint Operations.* 28 September 1998.

JP 2-03. *Joint Tactics, Techniques, and Procedures for Geospatial Information and Services Support to Joint Operations.* 31 March 1999.

JP 3-0. *Doctrine and Joint Operations.* 1 February 1995.

JP 3-07. *Joint Doctrine for Military Operations Other Than War.* 16 June 1995.

JP 3-07.2. *JTTP for Antiterrorism.* 17 March 1998.

JP 3-08. *Interagency Coordination During Joint Operations,* Volume II. 9 October 1996.

JP 3-09. *Doctrine for Joint Fire Support.* 12 May 1998.

JP 3-09.1. *Joint Tactics, Techniques and Procedures for Laser Designation Operations.* 28 May 1999.

JP 3-11. *Joint Doctrine for Operations in Nuclear, Biological, and Chemical (NBC) Environments.* 11 July 2000.

JP 3-13. *Joint Doctrine for Information Operations.* 9 October 1998.

JP 3-18. *Joint Doctrine for Forcible Entry Operations.* 16 July 2001.

JP 3-35. *Joint Deployment and Redeployment Operations.* 7 September 1999.

JP 3-51. *Joint Doctrine for Electronic Warfare.* 7 April 2000.

JP 3-55. *Doctrine for Reconnaissance, Surveillance, and Target Acquisition Support for Joint Operations (RSTA).* 14 April 1993.

JP 3-57. *Joint Doctrine for Civil-Military Operations.* 8 February 2001.

JP 3-60. *Joint Doctrine for Targeting.* 17 January 2002.

JP 4-0. *Doctrine for Logistic Support of Joint Operations.* 6 April 2000.

JP 5-0. *Doctrine for Planning Joint Operations.* 13 April 1995.

JP 5-00.2. *Joint Task Force Planning Guidance and Procedures.* 13 January 1999.

PUBLIC LAWS AND OTHER PUBLICATIONS

The United States Code is available online: http://uscode.house.gov/usc.htm

EO 12333. *United States Intelligence Activities.* 4 December 1981.

CJCS Instruction 1301.01, Policy and Procedures to Assign Individuals to Meet Combatant Command Mission-Related Temporary Duty Requirements. 1 July 2001.

CJCSM 3500.04C. *Universal Joint Task List.* 1 July 2002.

Military Standard 2525B, *DOD Interface Standard, Common Warfighting Symbology.* 30 January 1999.

PUBLICATIONS UNDER DEVELOPMENT

FM 2-01. *Intelligence Synchronization.* When published tentatively in July 2004, FM 2-01 will supersede FM 34-2, 8 March 1994, and FM 34-2-1, 19 June 1991.

FM 2-01.2. *Counterintelligence Analysis.* When published tentatively in April 2005, FM 2-01.2 will supersede FM 34-60, 3 October 1995.

FM 2-01.3. *Intelligence Preparation of the Battlefield.* When published tentatively December 2004, FM 2-01.3 will supersede FM 34-130, 8 July 1994.

Index

Entries are by paragraph number.

A

All-Source Intelligence. *See also* Intelligence Disciplines.
    Definition, 1-85, 5-1
    Fundamentals, 5-3
    Products, 5-4
        ECOA sketches, 5-4
        Event templates, 5-4
        Databases, 5-4
        HPT lists, 5-4
        Intelligence estimate, 5-16
        MCOO, 5-4
        Running estimate, 5-26
    Role, 5-2
Analysis, 1-3, 1-13, 1-17, 1-26, 1-33, 1-53, 1-56, 1-60, 1-67, 1-81, 1-89, 1-133, 6-10, 6-32
    All-source, 1-56, 2-36, 2-60, 5-2, 5-23, 5-32, 5-33, 6-9
    BOS, 4-6
    CI, 1-97, 2-26, 11-14, 11-42, 11-58, 11-59
    COA, 1-14, 1-33, 5-16
    Collection, 5-21
    COP, 5-27
    HUMINT, 6-49, 6-56
    I&W, 2-19
    Imagery, 7-4, 7-12, 7-14, 7-18, 7-19
    In AO, 2-11, 3-9, 3-15
    In support of FP, 1-16
    IPB, 5-10, 6-10
    MASINT, 9-4, 9-6, 9-18, 9-21
    Mission, 1-70, 6-20, 11-30, 11-61, B-3, B-7
    NGIC, 10-9
    Operational, 5-35, 6-10, 6-14, 11-14
    Predictive, 1-13, 2-6, 4-38, 5-12
    Process, 4-26, 4-28, 4-34, 4-35, 5-22, 8-26
    Requirements management, 4-36
    S&T, 2-24
    Strategic, 6-10, 11-14
    Target development, 2-23
    Teams, 2-59
    TECHINT, 10-1, 10-21
Analyze. *See* Intelligence Process.
Armed Forces Medical Intelligence Center (AFMIC), 10-6
Army DCS, 10-8
Assess. *See* Intelligence Process.
Area Studies of Foreign Countries, 1-20, 1-24, Figure 1-1. *See also* Intelligence Tasks; Support to Strategic Responsiveness.
Army capabilities, 3-17
    Combined arms, 3-19
    Command and support relationships, 3-20, Figure 3-1
ASCOPE. *See* Civil Considerations.

C

Categories of Intelligence, 2-17

Central Imagery Tasking Office (CITO), 7-9
Civil Considerations, 1-70
Combat assessment, 1-52 through 1-57, 1-60, 1-65, 1-105, Figure 1-1
Combat power
    Elements of, 3-11
Combat service support. *See* Battlefield Operating System.
Command and control. *See* Battlefield Operating System.
Common tasks. *See also* Intelligence Tasks.
    Analyze, 1-82, 4-35
    Assess, 1-84
    Disseminate, 1-83, 4-39
Conduct ISR. *See also* Intelligence Tasks.
    Conduct Surveillance, 1-26, 1-50, 1-51,
    Conduct Tactical Reconnaissance, 1-26, 1-48, 1-49
    Perform Intelligence Synchronization, 1-26 through 1-40
    Perform ISR integration, 1-26, 1-41 through 1-47
Counterintelligence. *See also* Intelligence Disciplines.
    Category of Intelligence, 2-26
    Command and control relationship, 11-53
    Definition, 1-97, 11-1
    Equipment, 11-37

Functions, 11-4 through 11-15
Levels of employment, 11-22
Operational employment, 11-21
Role, 11-2
Staff responsibilities, 11-56 through 11-62
Support to contingency operations, 11-25
Support to installations and operating bases, 11-29
Teams, 11-29 through 11-52, 11-61, 11-62
Counterintelligence Coordinating Authority (CICA), 11-57
Current Intelligence
   Categories of Intelligence, 2-20

D-E-F

Department of Energy, 2-38
Department of Homeland Security Mission, 2-41
Department of Justice, 2-40
Disseminate. *See* Intelligence Process; common tasks.
Elements of IO. *See* information operations.
Essential Elements of Friendly Information, 1-32, 1-38, Figure 1-4
Force projection operations, 1-107 through 1-138
   Deployment, 1-107, 1-120 through 1-128
   Employment, 1-107, 1-129 through 1-134
   Mobilization, 1-107, 1-114 through 1-119
   Redeployment, 1-107, 1-136 through 1-138
   Sustainment, 1-107, 1-135
Full spectrum operations, 3-1
   Defensive operations, 3-5
   Offensive operations, 3-4
   Stability operations, 3-8
   Support operations, 3-10
Functional damage assessment (FDA), 1-56

G-H

G2/S2. *See* Responsibilities.
General Military Intelligence
   Category of Intelligence, 2-21
Geospatial intelligence, 1-88
Human Intelligence (HUMINT). *See also* Intelligence Disciplines.
   Collection methods, 6-4
   Combat patrolling, 6-7
   Definition, 1-87, 6-1
   Equipment, 6-27
   Functions, 6-3
   ISR operations, 6-8
   Products, 6-11
   Role, 6-2
   Staff responsibilities, 6-47 through 6-55
   Support to contingency operations, 6-16
   Support to installations and operating bases, 6-19
   Strategic and departmental employment, 6-13
   Tactical Questioning, 6-6
   Teams, 6-20 through 6-43, 6-56, 6-57

I

Imagery Intelligence (IMINT). *See also* Intelligence Disciplines.
   Definition, 1-90, 7-1
   Functions, 7-11, 7-14 through 7-22
   Fundamentals, 7-3
   Role, 7-2, 7-13
   Sources of imagery, 7-5
   Types of imagery sensors, 7-10, 7-15, Table 7-1
Indications and Warnings, 1-20 through 1-22, Figure 1-1
   Category of Intelligence, 2-19
Information Operations, A-1
   Definition, A-1, A-5
   Elements of, A-2, A-9
   Intelligence support to, A-10
Information requirements, 1-28
   Develop, 1-30 through 1-36
   Friendly force, 1-37
Information Superiority, A-3
INSCOM. *See* Responsibilities.
Intelligence. *See also* intelligence BOS; intelligence warfighting function.
   Categories of, 2-17
   In full spectrum operations, 3-1
   role of, 1-1
Intelligence BOS. *See also* intelligence warfighting function.
   and the COP, 5-29
   considerations, 4-6
   human sensor information, 6-47, 11-56
   in full spectrum operations, 3-2
   in offensive operations, A-12
   in defensive operations, A-12
   interoperability, 6-30, 11-40, 11-56
   limitations, 4-6
   requirements, 4-47
Intelligence Community organizations, 2-27, Figure 2-2

FM 2-0, C1

Intelligence Disciplines, 1-85 through 1-97
　All-source Intelligence, 1-86, 5-1, 5-2 through 5-4, 5-16, 5-26
　CI, 1-97, 2-26, 11-1 through 11-52, 11-61, 11-62
　GEOINT, 1-88, 1-89
　HUMINT, 1-87, 6-1 through 6-3, 6-4 through 6-8, 6-11, 6-13, 6-19 through 6-43, 6-56, 6-57
　IMINT, 1-90, 7-1 through 7-5, 7-11 through 7-22, Table 7-1
　MASINT, 1-92, 9-1, 9-4, 9-5, 9-9 through 9-24
　OSINT, 1-93 through 1-95
　SIGINT, 1-91, 8-1, 8-3 through 8-5, 8-13, 8-14 through 8-34
　TECHINT, 1-96, 10-1, 10-6 through 10-10, 10-12 through 10-25
Intelligence estimates, 5-5
Intelligence Preparation of the Battlefield. See Intelligence Tasks; Support to Situational Understanding.
Intelligence Process, 1-72 through 1-84, 4-1, 4-2, Figure 4-2
　Collect, 1-76, 4-18, 5-21, 7-15, 8-22, 9-16, 10-16
　Common tasks, 1-81 through 1-84, 4-2
　　Analyze, 1-82, 4-35, 5-32, 6-10, 7-18, 8-30, 9-21, 10-22
　　Assess, 1-84, 4-48, 5-35, 7-22, 8-34, 9-24, 10-25
　　Disseminate, 1-83, 4-39, 5-33, 7-21, 8-32, 9-22, 10-23

Functions, 4-2
Plan, 1-72, 4-1, 4-4, 4-8, 5-4, 5-19, 7-11, 8-14, 9-12, 10-12
Prepare, 1-74, 1-75, 4-8, 5-15, 7-14, 8-18, 9-15, 10-15
Process, 1-77, 1-78, 4-28, 5-22, 7-16, 8-26, 9-18, 10-18
Produce, 1-79, 1-80, 4-31, 5-23, 7-17, 8-28, 9-19, 10-19
Intelligence Reach, 2-69
　Components, 2-72
　　Broadcast Services, 2-80
　　Collaborative Tools, 2-82, Table 2-2
　　Database Access, 2-80
　　Pull, 2-73
　　Push, 2-73
　　Requirements Management, 2-83
　　Partners and sources, Table 2-1
Intelligence Readiness, 1-9, 1-20, 1-23, 1-108, 1-109, 1-138, Figure 1-1
Intelligence support to IO, 1-52, 1-54, Figure 1-1
Intelligence support to targeting, 1-52, 1-53, Figure 1-1
Intelligence synchronization, 1-27 through 1-40
Intelligence synchronization plan, 1-28, 1-39, 1-40, 1-46, 1-61, 1-73, 1-116, 1-128, 2-53
Development, 1-39, 1-40, 5-51
Intelligence Support to Force Protection, 1-12, 1-16 through 1-18, Figure 1-1
Intelligence Tasks, 1-11 through 1-65

Support to Situational Understanding, 1-5, 1-12, through 1-19
Support to Strategic Responsiveness, 1-5, 1-20 through 1-25
Conduct Intelligence, Surveillance, and Reconnaissance, 1-5, 1-26 through 1-51
Provide Intelligence Support to Effects, 1-5, 1-52 through 1-65
Intelligence warfighting function, 1-5 through 1-13, 1-23, 1-35, 1-53, 1-64, 1-89, 1-109, 1-114
　Definition, 1-5
ISR Integration, 1-41 through 1-47, Figure 1-1
ISR Tasks, 4-19
ISR Plan, 1-36, 1-40, 1-41, 1-43 through 1-47, 1-53, 1-61, 1-62, 1-73, 4-40, 6-22, 6-56

J-L

Joint intelligence
　Architecture, 2-55
　Augmentation considerations, 2-62
　Considerations in, 2-52
　Operations, 2-48
　Organizations, 2-61
Joint Intelligence Center (JIC), 2-57
Joint Task Force J2 Organization, Figure 2-3
Levels of war, 2-3, Figure 2-1
　Operational, 2-9
　Strategic, 2-5
　Tactical, 2-15
Linguist Support, B-1
　Categories, B-1, B-8
　Proficiency, B-22, B-26

Responsibilities
  Personal Staff Officer, B-11
  Special Staff Officer, B-10
  Primary Staff, B-9
Requirements, B-3
Sources, B-12

M-O

Measurement and Signatures Intelligence (MASINT). *See also* Intelligence Disciplines.
  Definition, 9-1
  Functions, 9-12 through 9-24
  Fundamentals, 9-9
  Role, 9-5
  Subdisciplines, 9-4
METT-TC. *See* Mission Variables.
Mission Variables (METT-TC), 1-70, 1-71
Munitions effects assessment (MEA), 1-57, 1-58
Nature of Land Operations, 1-100 through 1-106
Office of Nonproliferation and National Security, 2-38
Open-Source Intelligence (OSINT), 1-93 through 1-95
Operational Environment, 1-66 through 1-71
Operational framework, 3-13
  Area of operations, 3-13
  Battlefield organization, 3-13
  Battlespace, 3-13
Operational Variables (PMESII-PT), 1-67 through 1-69
Operations officer. *See* Responsibilities.

P

Physical damage assessment (PDA), 1-56
Plan. *See* Intelligence Process.

PMESII-PT. *See* Operational Variables.
Police Intelligence Operations (PIO), 1-19, Figure 1-1
Predictive intelligence, 1-3, 1-17, 1-130 through 1-134
Prepare. *See* Intelligence Process.
Principles of war, 3-12
Priority intelligence requirements, 1-33 through 1-36, 1-38, 1-61, Figures 1-3 and 1-4
Process. *See* Intelligence Process.
Produce. *See* Intelligence Process.
Provide Intelligence Support to Effects. *See also* Intelligence Tasks.
  Provide Intelligence Support to Combat Assessment, 1-52, 1-55 through 1-65
  Provide Intelligence Support to Information Operations, 1-52, 1-54
  Provide Intelligence Support to Targeting, 1-52, 1-53

R-S

Responsibilities
  *2X staff, 6-47
  AFMIC, 10-6
  Army DCS, 10-8
  CICA, 11-57
  CITO, 7-9
  Commander's, 3-5, 3-13, 3-15, 3-21, 6-45, 11-54
  Department of Justice, 2-40
  Fire support personnel, 1-57
  G2/S2, 1-13, 2-19, 2-20, 3-16, 10-14
  G3/S3 (operations officer), 1-42 through 1-47
  INSCOM, 2-13, 9-10, 10-9

JIC, 2-57
Office of Nonproliferation & National Security, 2-38
Staff sections, B-9
RSTA/ISR, 1-105, 1-106
Running estimate, 5-26, 5-30
S&T Intelligence
  Category of Intelligence, 2-24
Sensitive Site Exploitation, 1-20, 1-25, Figure 1-1
Signals Intelligence (SIGINT), 1-91, 8-1. *See also* Intelligence Disciplines.
  at Echelons Corps and Below, 8-13
  Definition, 8-1
  Functions, 8-14 through 8-34
  Fundamentals, 8-4
  Role, 8-3
  TCAE Architecture, 8-5
Situation Development, 1-12, 1-14, 1-15, Figure 1-1. *See also* Intelligence tasks; support to situational understanding.
Situational Understanding, 1-12, Figure 1-1
Staff sections. *See* Responsibilities.
Support to force generation, 1-5
Support to Situational Understanding. *See also* Intelligence Tasks, 1-12 through 1-19, Figure 1-1
  Perform IPB, 1-12, 1-13, Figure 1-1
  Perform Situation Development 1-12, 1-14, 1-15, Figure 1-1
  Provide Intelligence Support to Force Protection, 1-12, 1-16 through 1-18, Figure 1-1

Conduct Police Intelligence Operations, 1-12, 1-19, Figure 1-1

Support to Strategic Responsiveness. *See also* Intelligence Tasks, 1-20 through 1-25, Figure 1-1

    Perform I&W, 1-20 through 1-22, Figure 1-1

    Ensure Intelligence Readiness, 1-20, 1-23, Figure 1-1

    Conduct Area Studies of Foreign Countries, 1-20, 1-24, Figure 1-1

    Support Sensitive Site Exploitation, 1-20, 1-25, Figure 1-1

T

Tactical Reconnaissance, 1-48, 1-49, Figure 1-1. *See also* Intelligence Tasks; Conduct ISR.

Tactical questioning. *See* Human Intelligence.

Target Intelligence

    Category of Intelligence, 2-23

Target system assessment (TSA), 1-56

Task organization, 3-18

Technical Intelligence (TECHINT). *See also* Intelligence Disciplines.

    Assets, 10-6 through 10-10

    Definition, 1-96, 10-1

    Functions, 10-12 through 10-25

    Fundamentals, 10-5

Tenets of Army operations, 3-12

U

Unified Action, 1-98 through 1-100

    Intelligence and, 2-1

    Intelligence operations, 2-11, 2-44

    Joint intelligence operations, 2-48

    Architecture, 2-15

    Augmentation considerations, 2-62

    Considerations in, 2-52

    JTF organizations, 2-61

    Multinational Intelligence, 2-66

W

Warfighting functions, 1-6

FM 3-36

Field Manual
No. 3-36

Headquarters
Department of the Army
Washington, DC, 25 February 2009

Electronic Warfare in Operations

Contents

	PREFACE iv
Chapter 1	ELECTRONIC WARFARE OVERVIEW 1-1
	Operational Environments 1-1
	Information and the Electromagnetic Spectrum 1-1
	Divisions of Electronic Warfare 1-4
	Activities and Terminology 1-7
	Summary 1-12
Chapter 2	ELECTRONIC WARFARE IN FULL SPECTRUM OPERATIONS 2-1
	The Role of Electronic Warfare 2-1
	The Application of Electronic Warfare 2-3
	Summary 2-7
Chapter 3	ELECTRONIC WARFARE ORGANIZATION 3-1
	Organizing Electronic Warfare Operations 3-1
	Planning and Coordinating Electronic Warfare Activities 3-4
	Summary 3-6
Chapter 4	ELECTRONIC WARFARE AND THE OPERATIONS PROCESS 4-1
	Section I — Electronic Warfare Planning 4-1
	The Military Decisionmaking Process 4-2
	Decisionmaking in a Time-Constrained Environment 4-9
	The Integrating Processes and Continuing Activities 4-10
	Employment Considerations 4-15
	Section II — Electronic Warfare Preparation 4-19
	Section III — Electronic Warfare Execution 4-19
	Section IV — Electronic Warfare Assessment 4-20
	Summary 4-21
Chapter 5	COORDINATION, DECONFLICTION, AND SYNCHRONIZATION 5-1
	Coordination and Deconfliction 5-1
	Synchronization 5-5
	Summary 5-5

Distribution Restriction: Approved for public release; distribution is unlimited.

Contents

Chapter 6	INTEGRATION WITH JOINT AND MULTINATIONAL OPERATIONS............. 6-1
	Joint Electronic Warfare Operations 6-1
	Multinational Electronic Warfare Operations 6-4
	Summary.. 6-6
Chapter 7	ELECTRONIC WARFARE CAPABILITIES 7-1
	Service Electronic Warfare Capabilities... 7-1
	External Support Agencies and Activities 7-1
	Summary.. 7-3
Appendix A	THE ELECTROMAGNETIC ENVIRONMENT... A-1
Appendix B	ELECTRONIC WARFARE INPUT TO OPERATION PLANS AND ORDERS. B-1
Appendix C	ELECTRONIC WARFARE RUNNING ESTIMATE.. C-1
Appendix D	ELECTRONIC WARFARE-RELATED REPORTS AND MESSAGES............. D-1
Appendix E	ARMY AND JOINT ELECTRONIC WARFARE CAPABILITIES...................... E-1
Appendix F	TOOLS AND RESOURCES RELATED TO ELECTRONIC WARFARE........... F-1
	GLOSSARY... Glossary-1
	REFERENCES.. References-1
	INDEX Index-1

Figures

Figure 1-1. The electromagnetic spectrum 1-2
Figure 1-2. Electromagnetic spectrum targets 1-3
Figure 1-3. The three subdivisions of electronic warfare 1-4
Figure 1-4. Means versus effects 1-12
Figure 2-1. Electronic warfare weight of effort during operations 2-2
Figure 3-1. Electronic warfare coordination organizational framework 3-2
Figure 4-1. The operations process 4-1
Figure 4-2. Example of analysis for an enemy center of gravity 4-3
Figure 4-3. Course of action development 4-5
Figure 4-4. Course of action comparison 4-8
Figure 4-5. Integrating processes and continuing activities... 4-10
Figure 4-6. Electronic warfare support to intelligence preparation of the battlefield 4-11
Figure 4-7. Electronic warfare in the targeting process 4-13
Figure 5-1. Spectrum deconfliction procedures 5-3
Figure 6-1. Joint frequency management coordination 6-3
Figure 6-2. Electronic warfare support request coordination... 6-4
Figure A-1. The electromagnetic spectrum.. A-2
Figure B-1. Appendix 4 (Electronic Warfare) to annex P (Information Operations) instructions B-2
Figure C-1. Example of an electronic warfare running estimate C-2

Preface

PURPOSE

FM 3-36 provides Army doctrine for electronic warfare (EW) planning, preparation, execution, and assessment in support of full spectrum operations. Users of FM 3-36 must be familiar with full spectrum operations established in FM 3-0; the military decisionmaking process established in FM 5-0; the operations process established in FMI 5-0.1; commander's visualization described in FM 6-0; and electronic warfare described in JP 3-13.1.

SCOPE

FM 3-36 is organized into seven chapters and six appendixes. Each chapter addresses a major aspect of Army EW operations. The appendixes address aspects of EW operations that complement the operational doctrine. A glossary contains selected terms.

- Chapter 1 discusses the nature and scope of electronic warfare and the impact of the electromagnetic environment on Army operations.
- Chapter 2 offers a discussion of EW support to full spectrum operations, combat power, the warfighting functions, and information tasks.
- Chapter 3 introduces the organizational framework for command and control of EW operations.
- Chapter 4 describes how commanders integrate EW operations throughout the operations process.
- Chapter 5 discusses the coordination required to synchronize and deconflict EW operations effectively.
- Chapter 6 provides the baseline for integrating EW operations into joint and multinational operations.
- Chapter 7 discusses the enabling activities that support EW operations, such as command and control, intelligence, logistics, technical support and EW training.
- Appendix A discusses the electromagnetic environment.
- Appendix B illustrates an EW appendix to an operation order.
- Appendix C illustrates an EW running estimate.
- Appendix D discusses EW related reports and messages.
- Appendix E offers a reference guide to Army and joint EW capabilities.
- Appendix F discusses EW-related tools and resources.

APPLICABILITY

FM 3-36 provides guidance on EW operations for commanders and staffs at all echelons. This FM serves as an authoritative reference for personnel who—

- Develop doctrine (fundamental principles and tactics, techniques, and procedures), materiel, and force structure.
- Develop institutional and unit training.
- Develop standing operating procedures for unit operations.
- Conduct planning, preparation, execution and assessment of electronic warfare.

FM 3-36 applies to the Active Army, Army National Guard/Army National Guard of the United States, and U.S. Army Reserve, unless otherwise stated.

Chapter 1
Electronic Warfare Overview

This chapter provides an overview of electronic warfare and the conceptual foundation that leaders require to understand the electromagnetic environment and its impact on Army operations.

OPERATIONAL ENVIRONMENTS

1-1. An *operational environment* is a composite of the conditions, circumstances, and influences that affect the employment of capabilities and bear on the decisions of the commander (JP 3-0). An operational environment includes physical areas—the air, land, maritime, and space domains. It also includes the information that shapes the operational environment as well as enemy, adversary, friendly, and neutral systems relevant to a joint operation. Joint planners analyze operational environments in terms of six interrelated operational variables: political, military, economic, social, information, and infrastructure. To these variables Army doctrine adds two more: physical environment and time. (See FM 3-0 for additional information on the operational variables). Army leaders use operational variables to understand and analyze the broad environment in which they are conducting operations.

1-2. Army leaders use mission variables to synthesize operational variables and tactical-level information with local knowledge about conditions relevant to their mission. They use mission variables to focus analysis on specific elements that directly affect their mission. Upon receipt of a warning order or mission, Army tactical leaders narrow their focus to six mission variables known as METT-TC. They are mission, enemy, terrain and weather, troops and support available, time available and civil considerations. The mission variables outline the situation as it applies to a specific Army unit.

1-3. Commanders employ and integrate their unit's capabilities and actions within their operational environment to achieve a desired end state. Through analyzing their operational environment, commanders understand how the results of friendly, adversary, and neutral actions may impact that end state. During military operations, both friendly and enemy commanders depend on the flow of information to make informed decisions. This flow of information depends on the electronic systems and devices used to communicate, navigate, sense, store, and process information.

INFORMATION AND THE ELECTROMAGNETIC SPECTRUM

1-4. Commanders plan for and operate electronic systems and the weapon systems that depend on them in an intensive and nonpermissive electromagnetic environment. They ensure the flow of information required for their decisionmaking. (Appendix A further discusses the electromagnetic environment.) Within the electromagnetic environment, electronic systems and devices operate in the electromagnetic spectrum. (See figure 1-1, page 1-2.)

1-5. The electromagnetic spectrum has been used for commercial and military applications for over a century. However, the full potential for its use as the primary enabler of military operations is not yet fully appreciated. New technologies are expanding beyond the traditional radio frequency spectrum. They include high-power microwaves and directed-energy weapons. These new technologies are part of an electronic warfare (EW) revolution by military forces. Just as friendly forces leverage the electromagnetic spectrum to their advantage, so do capable enemies use the electromagnetic spectrum to threaten friendly force operations. The threat is compounded by the growth of a wireless world and the increasingly sophisticated use of commercial off-the-shelf technologies.

Chapter 1

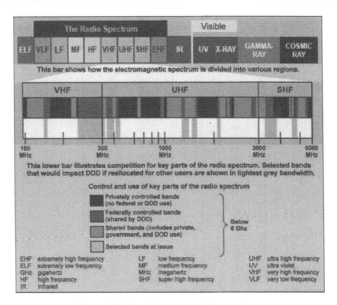

Figure 1-1. The electromagnetic spectrum

1-6. Adversaries and enemies, from small and single actors to large state, multinational, and nonstate actors, use the most modern technology. Such technology is moving into the cellular and satellite communications area. Most military and commercial operations rely on electromagnetic technologies and are susceptible to the inherent vulnerabilities associated with their use. This reliance requires Army forces to dominate the electromagnetic spectrum (within their operational environment) with the same authority that they dominate traditional land warfare operations. Emerging electromagnetic technologies offer expanded EW capabilities. They dynamically affect the electromagnetic spectrum through delivery and integration with other types of emerging weapons and capabilities. Examples are directed-energy weapons, high-powered microwaves, lasers, infrared, and electro-optical and wireless networks and devices.

Electronic Warfare Overview

1-7. In any conflict, commanders attempt to dominate the electromagnetic spectrum. They do this by locating, targeting, exploiting, disrupting, degrading, deceiving, denying, or destroying the enemy's electronic systems that support military operations or deny the spectrum's use by friendly forces. The increasing portability and affordability of sophisticated electronic equipment guarantees that the electromagnetic environment in which forces operate will become even more complex. To ensure unimpeded access to and use of the electromagnetic spectrum, commanders plan, prepare, execute, and assess EW operations against a broad set of targets within the electromagnetic spectrum. (See figure 1-2.)

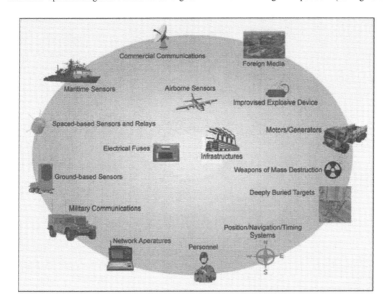

Figure 1-2. Electromagnetic spectrum targets

25 February 2009　　　FM 3-36　　　1-3

Chapter 1

DIVISIONS OF ELECTRONIC WARFARE

1-8. *Electronic warfare* is defined as military action involving the use of electromagnetic and directed energy to control the electromagnetic spectrum or to attack the enemy. Electronic warfare consists of three divisions: electronic attack, electronic protection, and electronic warfare support (JP 3-13.1). (See figure 1-3.)

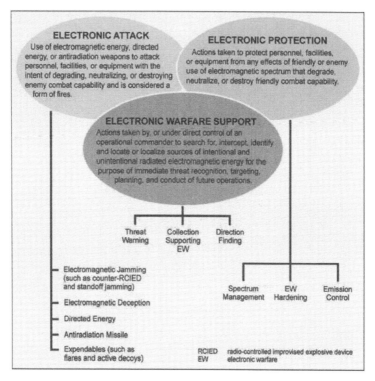

Figure 1-3. The three subdivisions of electronic warfare

Electronic Attack

1-9. *Electronic attack* is a division of electronic warfare involving the use of electromagnetic energy, directed energy, or antiradiation weapons to attack personnel, facilities, or equipment with the intent of degrading, neutralizing, or destroying enemy combat capability and is considered a form of fires (JP 3-13.1). Electronic attack includes—

- Actions taken to prevent or reduce an enemy's effective use of the electromagnetic spectrum, such as jamming and electromagnetic deception.
- Employment of weapons that use either electromagnetic or directed energy as their primary destructive mechanism (lasers, radio frequency weapons, particle beams).
- Offensive and defensive activities including countermeasures.

1-10. Common types of electronic attack include spot, barrage, and sweep electromagnetic jamming. Electronic attack actions also include various electromagnetic deception techniques such as false target or duplicate target generation. (See paragraphs 1-23 to 1-31 for further discussion of electronic attack activities.)

1-11. *Directed energy* is an umbrella term covering technologies that relate to the production of a beam of concentrated electromagnetic energy or atomic or subatomic particles (JP 1-02). A directed-energy weapon uses directed energy primarily as a direct means to damage or destroy an enemy's equipment, facilities, and personnel. In addition to destructive effects, directed-energy weapon systems support area denial and crowd control. (See appendix A for more information on directed energy.)

1-12. Examples of offensive electronic attack include—

- Jamming enemy radar or electronic command and control systems.
- Using antiradiation missiles to suppress enemy air defenses (antiradiation weapons use radiated energy emitted from the target as their mechanism for guidance onto targeted emitters).
- Using electronic deception techniques to confuse enemy intelligence, surveillance, and reconnaissance systems.
- Using directed-energy weapons to disable an enemy's equipment or capability.

1-13. Defensive electronic attack uses the electromagnetic spectrum to protect personnel, facilities, capabilities, and equipment. Examples include self-protection and other protection measures such as use of expendables (flares and active decoys), jammers, towed decoys, directed-energy infrared countermeasure systems, and counter-radio-controlled improvised-explosive-device systems. (See JP 3-13.1 for more discussion of electronic attack.)

Electronic Protection

1-14. *Electronic protection* is a division of electronic warfare involving actions taken to protect personnel, facilities, and equipment from any effects of friendly or enemy use of the electromagnetic spectrum that degrade, neutralize, or destroy friendly combat capability (JP 3-13.1). For example, electronic protection includes actions taken to ensure friendly use of the electromagnetic spectrum, such as frequency agility in a radio, or variable pulse repetition frequency in radar. Electronic protection should not be confused with self-protection. Both defensive electronic attack and electronic protection protect personnel, facilities, capabilities, and equipment. However, electronic protection protects from the effects of electronic attack (friendly and enemy), while defensive electronic attack primarily protects against lethal attacks by denying enemy use of the electromagnetic spectrum to guide or trigger weapons.

1-15. During operations, electronic protection includes, but is not limited to, the application of training and procedures for countering enemy electronic attack. Army commanders and forces understand the threat and vulnerability of friendly electronic equipment to enemy electronic attack and take appropriate actions to safeguard friendly combat capability from exploitation and attack. Electronic protection measures minimize the enemy's ability to conduct electronic warfare support (electronic warfare support is discussed in paragraphs 1-18 to 1-20) and electronic attack operations successfully against friendly forces. To protect friendly combat capabilities, units—

- Regularly brief force personnel on the EW threat.
- Ensure that electronic system capabilities are safeguarded during exercises, workups, and predeployment training.
- Coordinate and deconflict electromagnetic spectrum usage.
- Provide training during routine home station planning and training activities on appropriate electronic protection active and passive measures.
- Take appropriate actions to minimize the vulnerability of friendly receivers to enemy jamming (such as reduced power, brevity of transmissions, and directional antennas).

1-16. Electronic protection also includes spectrum management. The spectrum manager works for the G-6 or S-6 and plays a key role in the coordination and deconfliction of spectrum resources allocated to the force. Spectrum managers or their direct representatives participate in the planning for EW operations.

1-17. The development and acquisition of communications and electronic systems includes electronic protection requirements to clarify performance parameters. Army forces design their equipment to limit inherent vulnerabilities. If electronic attack vulnerabilities are detected, then units must review these programs. (See DODI 4650.01 for information on the spectrum certification process and electromagnetic compatibility.)

ELECTRONIC WARFARE SUPPORT

1-18. *Electronic warfare support* is a division of electronic warfare involving actions tasked by, or under the direct control of, an operational commander to search for, intercept, identify, and locate or localize sources of intentional and unintentional radiated electromagnetic energy for the purpose of immediate threat recognition, targeting, planning, and conduct of future operations (JP 3-13.1).

1-19. Electronic warfare support systems are a source of information for immediate decisions involving electronic attack, electronic protection, avoidance, targeting, and other tactical employments of forces. Electronic warfare support systems collect data and produce information or intelligence to—

- Corroborate other sources of information or intelligence.
- Conduct or direct electronic attack operations.
- Initiate self-protection measures.
- Task weapon systems.
- Support electronic protection efforts.
- Create or update EW databases.
- Support information tasks.

1-20. Electronic warfare support and signals intelligence missions use the same resources. The two differ in the detected information's intended use, the degree of analytical effort expended, the detail of information provided, and the time lines required. Like tactical signals intelligence, electronic warfare support missions respond to the immediate requirements of a tactical commander. Signals intelligence above the tactical level is under the operational control of the National Security Agency and directly supports the overarching national security mission. Resources that collect tactical-level electronic warfare support data can simultaneously collect national-level signals intelligence. See FM 2-0 for more information on signals intelligence.

ACTIVITIES AND TERMINOLOGY

1-21. Although new equipment and tactics, techniques, and procedures continue to be developed, the physics of electromagnetic energy remains constant. Hence, effective EW activities remain the same despite changes in hardware and tactics. Principal EW activities are discussed in the following paragraphs.

PRINCIPAL ACTIVITIES

1-22. Principal EW activities support full spectrum operations by exploiting the opportunities and vulnerabilities inherent in the use of the electromagnetic spectrum. The numerous EW activities are categorized by the EW subdivisions with which they are most closely associated: electronic attack, electronic warfare support, and electronic protection. JP 3-13.1 discusses these principal activities in detail.

Electronic Attack Activities

1-23. Activities related to electronic attack are either offensive or defensive and include—
- Countermeasures.
- Electromagnetic deception.
- Electromagnetic intrusion.
- Electromagnetic jamming.
- Electromagnetic pulse.
- Electronic probing.

Countermeasures

1-24. *Countermeasures* are that form of military science that, by the employment of devices and/or techniques, has as its objective the impairment of the operational effectiveness of enemy activity (JP 1-02). They can be deployed preemptively or reactively. Devices and techniques used for EW countermeasures include electro-optical-infrared countermeasures and radio frequency countermeasures.

1-25. *Electro-optical-infrared countermeasures* consist of any device or technique employing electro-optical-infrared materials or technology that is intended to impair or counter the effectiveness of enemy activity, particularly with respect to precision guided weapons and sensor systems. Electro-optical-infrared is the part of the electromagnetic spectrum between the high end of the far infrared and the low end of ultraviolet. Electro-optical-infrared countermeasures may use laser and broadband jammers, smokes/aerosols, signature suppressants, decoys, pyrotechnics/pyrophorics, high-energy lasers, or directed infrared energy countermeasures (JP 3-13.1).

1-26. *Radio frequency countermeasures* consist of any device or technique employing radio frequency materials or technology that is intended to impair the effectiveness of or counter enemy activity, particularly with respect to precision guided weapons and sensor systems (JP 3-13.1).

Electromagnetic Deception

1-27. *Electromagnetic deception* is the deliberate radiation, reradiation, alteration, suppression, absorption, denial, enhancement, or reflection of electromagnetic energy in a manner intended to convey misleading information to an enemy or to enemy electromagnetic-dependent weapons, thereby degrading or neutralizing the enemy's combat capability (JP 3-13.4). Among the types of electromagnetic deception are the following:
- Manipulative electromagnetic deception involves actions to eliminate revealing, or convey misleading, electromagnetic telltale indicators that may be used by hostile forces.
- Simulative electromagnetic deception involves actions to simulate friendly, notional, or actual capabilities to mislead hostile forces.
- Imitative electromagnetic deception introduces electromagnetic energy into enemy systems that imitates enemy emissions.

Chapter 1

Electromagnetic Intrusion

1-28. *Electromagnetic intrusion* is the intentional insertion of electromagnetic energy into transmission paths in any manner, with the objective of deceiving operators or of causing confusion (JP 1-02).

Electromagnetic Jamming

1-29. *Electromagnetic jamming* is the deliberate radiation, re-radiation, or reflection of electromagnetic energy for the purpose of preventing or reducing an enemy's effective use of the electromagnetic spectrum, with the intent of degrading or neutralizing the enemy's combat capability (JP 1-02).

Electromagnetic Pulse

1-30. *Electromagnetic pulse* is the electromagnetic radiation from a strong electronic pulse, most commonly caused by a nuclear explosion that may couple with electrical or electronic systems to produce damaging current and voltage surges (JP 1-02).

Electronic Probing

1-31. *Electronic probing* is the intentional radiation designed to be introduced into the devices or systems of potential enemies for the purpose of learning the functions and operational capabilities of the devices (JP 1-02). This activity is coordinated through joint or interagency channels and supported by Army forces.

Electronic Warfare Support Activities

1-32. Activities related to electronic warfare support include—
- Electronic reconnaissance.
- Electronic intelligence.
- Electronics security.

Electronic Reconnaisance

1-33. *Electronic reconnaissance* is the detection, location, identification, and evaluation of foreign electromagnetic radiations (JP 1-02).

Electronic Intelligence

1-34. *Electronic intelligence* is technical and geolocation intelligence derived from foreign noncommunications electromagnetic radiations emanating from other than nuclear detonations or radioactive sources (JP 1-02).

Electronics Security

1-35. *Electronics security* is the protection resulting from all measures designed to deny unauthorized persons information of value that might be derived from their interception and study of noncommunications electromagnetic radiations, e.g., radar (JP 1-02).

Electronic Protection Activities

1-36. Activities related to electronic protection include—
- Electromagnetic hardening.
- Electromagnetic interference.
- Electronic masking.
- Electronic warfare reprogramming.
- Emission control.
- Spectrum management.
- Wartime reserve modes.
- Electromagnetic compatibility.

Electromagnetic Hardening

1-37. *Electromagnetic hardening* consists of action taken to protect personnel, facilities, and/or equipment by filtering, attenuating, grounding, bonding, and/or shielding against undesirable effects of electromagnetic energy (JP 1-02).

Electromagnetic Interference

1-38. *Electromagnetic interference* is any electromagnetic disturbance that interrupts, obstructs, or otherwise degrades or limits the effective performance of electronics and electrical equipment. It can be induced intentionally, as in some forms of electronic warfare, or unintentionally, as a result of spurious emissions and responses, intermodulation products and the like (JP 1-02).

Electronic Masking

1-39. *Electronic masking* is the controlled radiation of electromagnetic energy on friendly frequencies in a manner to protect the emissions of friendly communications and electronic systems against enemy electronic warfare support measures/signals intelligence, without significantly degrading the operation of friendly systems (JP 1-02).

Electronic Warfare Reprogramming

1-40. *Electronic warfare reprogramming* is the deliberate alteration or modification of electronic warfare or target sensing systems, or the tactics and procedures that employ them, in response to validated changes in equipment, tactics, or the electromagnetic environment. These changes may be the result of deliberate actions on the part of friendly, adversary, or third parties; or may be brought about by electromagnetic interference or other inadvertent phenomena. The purpose of electronic warfare reprogramming is to maintain or enhance the effectiveness of electronic warfare and target sensing system equipment. Electronic warfare reprogramming includes changes to self-defense systems, offensive weapons systems, and intelligence collection systems (JP 3-13.1).

Emission Control

1-41. *Emission control* is the selective and controlled use of electromagnetic, acoustic, or other emitters to optimize command and control capabilities while minimizing transmissions for operations security: a. detection by enemy sensors; b. mutual interference among friendly systems; and/or c. enemy interference with the ability to execute a military deception plan (JP 1-02).

Chapter 1

Electromagnetic Spectrum Management

1-42. *Electromagnetic spectrum management* is planning, coordinating, and managing joint use of the electromagnetic spectrum through operational, engineering, and administrative procedures. The objective of spectrum management is to enable electronic systems to perform their functions in the intended environment without causing or suffering unacceptable interference (JP 6-0).

Wartime Reserve Modes

1-43. *Wartime reserve modes* are characteristics and operating procedures of sensors, communications, navigation aids, threat recognition, weapons, and countermeasures systems that will contribute to military effectiveness if unknown to or misunderstood by opposing commanders before they are used, but could be exploited or neutralized if known in advance. Wartime reserve modes are deliberately held in reserve for wartime or emergency use and seldom, if ever, applied or intercepted prior to such use (JP 1-02).

Electromagnetic Compatibility

1-44. *Electromagnetic compatibility* is the ability of systems, equipment, and devices that utilize the electromagnetic spectrum to operate in their intended operational environments without suffering unacceptable degradation or causing unintentional degradation because of electromagnetic radiation or response. It involves the application of sound electromagnetic spectrum management; system, equipment, and device design configuration that ensures interference-free operation; and clear concepts and doctrines that maximize operational effectiveness (JP 1-02).

APPLICATION TERMINOLOGY

1-45. EW capabilities are applied from the air, land, sea, and space by manned, unmanned, attended, or unattended systems. Units employ EW capabilities to achieve the desired lethal or nonlethal effect on a given target. Units maintain freedom of action in the electromagnetic spectrum while controlling the use of it by the enemy. Regardless of the application, units employing EW capabilities must use appropriate levels of control and protection of the electromagnetic spectrum. In this way, they avoid adversely affecting friendly forces. (Improper EW actions must be avoided because they may cause fratricide or eliminate high-value intelligence targets.)

1-46. In the context of EW application, units use several terms to facilitate control and protection of the electromagnetic spectrum. Terms used in EW application include control, detection, denial, deception, disruption and degradation, protection, and destruction. The three subdivisions of EW—electronic attack, electronic protection, and electronic warfare support—are specified within the following descriptions.

Control

1-47. In the context of EW, control of the electromagnetic spectrum is achieved by effectively coordinating friendly systems while countering enemy systems. Electronic attack limits enemy use of the electromagnetic spectrum. Electronic protection secures use of the electromagnetic spectrum for friendly forces, and electronic warfare support enables the commander's accurate assessment of the situation. All three are integrated for effectiveness. Commanders ensure maximum integration of communications; intelligence, surveillance, and reconnaissance; and information tasks.

Detection

1-48. In the context of EW, detection is the active and passive monitoring of the operational environment for radio frequency, electro-optic, laser, infrared, and ultraviolet electromagnetic threats. Detection is the first step in EW for exploitation, targeting, and defensive planning. Friendly forces maintain the capability to detect and characterize interference as hostile jamming or unintentional electromagnetic interference.

Denial

1-49. In the context of EW, denial is controlling the information an enemy receives via the electromagnetic spectrum and preventing the acquisition of accurate information about friendly forces. Degradation uses traditional jamming techniques, expendable countermeasures, destructive measures, or network applications. These range from limited effects up to complete denial of usage.

Deception

1-50. In the context of EW, deception is confusing or misleading an enemy by using some combination of human-produced, mechanical, or electronic means. Through use of the electromagnetic spectrum, EW deception manipulates the enemy's decision loop, making it difficult to establish accurate situational awareness.

Disruption and Degradation

1-51. In the context of EW, disruption and degradation techniques interfere with the enemy's use of the electromagnetic spectrum to limit enemy combat capabilities. This is achieved with electronic jamming, electronic deception, and electronic intrusion. These enhance attacks on hostile forces and act as force multipliers by increasing enemy uncertainty, while reducing uncertainty for friendly forces. Advanced electronic attack techniques offer the opportunity to nondestructively disrupt or degrade enemy infrastructure.

Protection

1-52. In the context of EW, protection is the use of physical properties; operational tactics, techniques, and procedures; and planning and employment processes to ensure friendly use of the electromagnetic spectrum. This includes ensuring that offensive EW activities do not electronically destroy or degrade friendly intelligence sensors or communications systems. Protection is achieved by component hardening, emission control, and frequency management and deconfliction. Frequency management and deconfliction include the capability to detect, characterize, geolocate, and mitigate electromagnetic interference that affects operations. Protection includes other means to counterattack and defeat enemy attempts to control the electromagnetic spectrum. Additionally, organizations such as a joint force commander's EW staff or a joint EW coordination cell enhance electronic protection by deconflicting EW efforts.

Destruction

1-53. Destruction, in the context of EW, is the elimination of targeted enemy systems. Sensors and command and control nodes are lucrative targets because their destruction strongly influences the enemy's perceptions and ability to coordinate actions. Various weapons and techniques ranging from conventional munitions and directed energy weapons to network attacks can destroy enemy systems that use the electromagnetic spectrum. Electronic warfare support provides target location and related information. While destroying enemy equipment can effectively deny the enemy use of the electromagnetic spectrum, the duration of denial will depend on the enemy's ability to reconstitute. (See JP 3-13.1.)

MEANS VERSUS EFFECTS

1-54. EW means are applied against targets to create a full range of lethal and nonlethal effects. (See figure 1-4.) Choosing a specific EW capability depends on the desired effect on the target and other considerations, such as time sensitivity or limiting collateral damage. EW capabilities provide commanders with additional options for achieving their objectives. During major combat operations there may be circumstances where commanders want to limit the physical damage on a given target. Under such circumstances, the EW staff articulates clearly to the commander the lethal and nonlethal effects EW capabilities can achieve. For example, a target might be enemy radar mounted on a fixed tower. Two EW options to defeat the radar could be to jam the radar or destroy it with antiradiation missiles. If the commander desired to limit damage to the tower, an electronic attack jamming platform would be preferred. In circumstances where commanders cannot sufficiently limit undesired effects such as collateral damage, they may be constrained from applying physical force. The EW staff articulates succinctly how EW capabilities can support actions to achieve desired effects and provide lethal and nonlethal options for commanders.

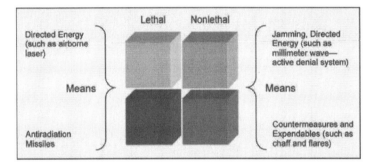

Figure 1-4. Means versus effects

SUMMARY

1-55. As the modern battlefield becomes more technologically sophisticated, military operations continue to be executed in an increasingly complex electromagnetic environment. Therefore, commanders and staffs need to thoroughly understand and articulate how the electromagnetic environment impacts their operations and how friendly EW operations can be used to gain an advantage. Commanders and staffs use the terminology presented in this chapter to describe the application of EW. This ensures a common understanding and consistency within plans, orders, standing operating procedures, and directives.

Chapter 2

Electronic Warfare in Full Spectrum Operations

Information technology is becoming universally available. Most enemies rely on communications and computer networks to make and implement decisions. Radios remain the backbone of tactical military command and control architectures. However, most communications relayed over radio networks are becoming digital as more computers link networks through transmitted frequencies. Therefore, the ability to dominate the electromagnetic spectrum is central to full spectrum operations. This chapter describes how commanders apply electronic warfare capabilities to support full spectrum operations.

THE ROLE OF ELECTRONIC WARFARE

2-1. Army electronic warfare (EW) operations seek to provide the land force commander with capabilities to support full spectrum operations. Full spectrum operations consist of the purposeful, simultaneous combination of offense, defense, and stability or civil support. The goal of full spectrum operations is to change the operational environment so that peaceful processes are dominant. Nonetheless, operational environments are complex; commanders must conduct operations across the entire spectrum of conflict. The Army maintains flexible forces with balanced capabilities and capacities. These flexible and balanced forces remain able to conduct major operations while executing other day-to-day smaller-scale operations. (See FM 3-0.)

2-2. Figure 2-1 (page 2-2) shows the weight of effort for using EW during operations. This figure adapts the elements of full spectrum operations (offense, defense, and stability or civil support) as described in FM 3-0. Overseas, Army forces conduct full spectrum operations (offensive, defensive, and stability) simultaneously as part of a joint force. Within the United States, Army forces conduct homeland defense and civil support operations as part of homeland security. Army electronic warfare (EW) operations seek to provide the land force commander with capabilities to support full spectrum operations. As noted in figure 2-1, statutory law limits the use of EW capabilities in support of civil support operations.

Chapter 2

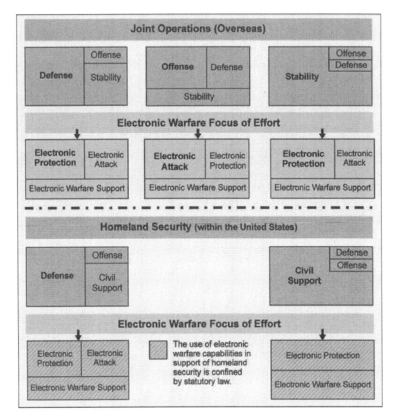

Figure 2-1. Electronic warfare weight of effort during operations

2-3. Full spectrum operations involve more than executing all elements of operations simultaneously. They require that commanders and staffs consider their unit's capabilities and capacities relative to each of the elements of full spectrum operations. Commanders consider how much can be accomplished simultaneously, how much can be phased, and what nonorganic resources may be available to solve problems. The same applies to EW in support of full spectrum operations. Commanders and staffs determine which resident and joint force EW capabilities to leverage in support of each element of full spectrum operations. Weighting the EW focus of effort within each of the elements assists commanders and their staffs in visualizing how EW capabilities can support their operations. Commanders combine offensive, defensive, and stability or civil support operations to seize, retain, and exploit the initiative. As they apply the appropriate level of EW effort to support these elements, commanders can seize, retain, and exploit the initiative within the electromagnetic environment.

THE APPLICATION OF ELECTRONIC WARFARE

2-4. To support full spectrum operations and achieve the goal of electromagnetic spectrum dominance, commanders fully integrate EW capabilities and apply them across the elements of combat power. Leadership and information are applied through, and multiply the effects of, the other six elements of combat power. Paragraphs 2-5 through 2-16 discuss the elements of combat power and how EW capabilities can support them.

IN SUPPORT OF LEADERSHIP

2-5. Leadership initiates the conditions for success. Commanders balance the ability to mass the effects of lethal and nonlethal systems with the requirements to deploy and sustain the units that employ those systems. Generating and maintaining combat power throughout an operation is essential. Today's operational environments require leaders who are competent, confident, and informed in using and protecting combat capabilities that operate within the electromagnetic spectrum. Commanders plan, prepare, execute, and assess EW operations to dominate the electromagnetic spectrum within their operational environment. To accomplish this domination, commanders effectively apply and integrate EW operations across the warfighting functions.

IN SUPPORT OF INFORMATION TASKS AND CAPABILITIES

2-6. Information is the element of combat power consisting of meaningful facts, data, and impressions used to develop a common situational understanding, to enable battle command, and to affect the operational environment. (See FM 3-0 for a discussion of combat power.) In modern conflict, gaining information superiority has become as important as lethal action in determining the outcome of operations. *Information superiority* is the operational advantage derived from the ability to collect, process, and disseminate an uninterrupted flow of information while exploiting or denying an adversary's ability to do the same (JP 3-13). To achieve this operational advantage, Army commanders direct efforts that contribute to information superiority. These efforts fall into four primary areas: Army information tasks; intelligence, surveillance, and reconnaissance; knowledge management; and information management. (See FM 3-0 for a discussion of information superiority.)

2-7. The Army information tasks are used to shape a commander's operational environment. These tasks are information engagement, command and control warfare, information protection, operations security, and military deception. Information capabilities can be used to produce both destructive and constructive effects. For example, destructive actions use information capabilities against the enemy's command and control system and other assets to reduce their combat capability. Constructive actions use information capabilities to inform or influence a particular audience or as a means to affect enemy morale. Although applicable to all elements of full spectrum operations, EW capabilities play a major role in enabling and supporting the execution of the command and control warfare and information protection tasks.

Chapter 2

2-8. *Command and control warfare* is the integrated use of physical attack, electronic warfare, and computer network operations, supported by intelligence, to degrade, destroy, and exploit an enemy's or adversary's command and control system or to deny information to it (FM 3-0). It includes operations intended to degrade, destroy, and exploit an enemy's or adversary's ability to use the electromagnetic spectrum and computer and telecommunications networks. *Information protection* is active or passive measures that protect and defend friendly information and information systems to ensure timely, accurate, and relevant friendly information. Information protection denies enemies, adversaries, and others the opportunity to exploit friendly information and information systems for their own purposes (FM 3-0). Table 2-1 shows capabilities, intended effects, staff responsibilities, and functional cells for the command and control warfare and information protection tasks. (For further information on the information tasks, refer to FM 3-0.)

Table 2-1. Two Army information tasks: command and control warfare and information protection

Army Information Tasks	Capabilities	Staff Responsibility	Functional Coordinating Cell	Intended Effects	Integrating Process
Command and Control Warfare	Physical Attack Electronic Attack Computer Network Attack Electronic Warfare Support Computer Network Exploitation	G-3/G-2	Fires	Degrade, disrupt, destroy and exploit enemy command and control	Operations Process
Information Protection	Information Assurance Computer Network Defense Electronic Protection	G-6	Network Operations	Protect friendly computer networks and communication means	

G-2 assistant chief of staff, intelligence
G-3 assistant chief of staff, operations
G-6 assistant chief of staff, signal

2-9. To support these information tasks, commanders ensure EW is coordinated, integrated, and synchronized with all other tasks. This occurs within the operations process through the various functional and integrating cells. Table 2-2 illustrates EW capabilities, actions, and objectives that support the command and control warfare and information protection tasks.

Table 2-2. Electronic warfare support to two Army information tasks

Information Tasks	Command and Control Warfare	Information Protection
Electronic warfare supports by	Locating and identifying threat command and control systems. Denying, disrupting, degrading, and/or destroying the enemy's command and control system. Supporting and complementing computer network attack and computer network exploitation operations.	Deconflicting spectrum usage with the spectrum manager. Hardening equipment against electromagnetic interference. Emissions control.
Action	Electronic attack (jamming, antiradiation missiles). Directed energy and electromagnetic spectrum area denial systems. Expendables (chaff, decoys, and flares). Electronic warfare support/signals intelligence.	Frequency agility in radios. Electronic shielding for systems. Electronic masking. Processes to counter intrusion. Implementing emissions control procedures to safeguard friendly systems and facilities from the effects of friendly and enemy electronic attack.
Objective or Effect	Detect, deny, disrupt or degrade, and destroy.	Control and protection.

IN SUPPORT OF THE WARFIGHTING FUNCTIONS

2-10. EW capabilities support each of the six warfighting functions. Examples of specific supporting capabilities are given in the following paragraphs.

Movement and Maneuver

2-11. The *movement and maneuver warfighting function* is the related tasks and systems that move forces to achieve a position of advantage in relation to the enemy. Direct fire is inherent in maneuver, as is close combat (FM 3-0). EW capabilities that enable the movement and maneuver of Army forces include—
- Suppression and destruction of enemy integrated air defenses.
- Denial of enemy information systems and intelligence, surveillance, and reconnaissance sensors.
- Target designation and range finding.
- Protection from effects of friendly and enemy EW.
- Lethal and nonlethal effects against enemy combat capability (personnel, facilities, and equipment).
- Threat warning and direction finding.
- Use of the electromagnetic spectrum to counter improvised explosive device operations.
- Electromagnetic spectrum obscuration, low observability, and multispectral stealth.

Intelligence

2-12. The *intelligence warfighting function* is the related tasks and systems that facilitate understanding of the operational environment, enemy, terrain, and civil considerations (FM 3-0). It includes tasks associated with intelligence, surveillance, and reconnaissance. EW capabilities that enable the intelligence warfighting function include—

- Increased access for intelligence collection assets (systems and personnel) by reducing antiaccess, antipersonnel, and antisystems threats.
- Increased capability to search for, intercept, identify, and locate sources of radiated electromagnetic energy in support of targeting, information tasks, and future operations.
- Increased capability in providing threat recognition and threat warning to the force.
- Indications and warning of threat emitters and radar.
- Denial and destruction of counter-intelligence, -surveillance, and -reconnaissance systems.

Fires

2-13. The *fires warfighting function* is the related tasks and systems that provide collective and coordinated use of Army indirect fires, joint fires, and command and control warfare, including nonlethal fires, through the targeting process (FM 3-0). It includes tasks associated with integrating command and control warfare. EW capabilities that enable the fires warfighting function include—

- Detection and location of targets radiating electromagnetic energy.
- Disruption, degradation, and destruction options for servicing targets. This includes information systems, targets requiring precision strike (such as minimal collateral damage and minimal weapons signature), hard and deeply buried targets, weapons of mass destruction, and power generation and infrastructure targets.
- Control, dispersion, or neutralization of combatant and noncombatant personnel with nonpersistent effects and minimum collateral damage (scalable and nonlethal).
- Area denial capabilities against vehicles, vessels, and aircraft.

Sustainment

2-14. The *sustainment warfighting function* is the related tasks and systems that provide support and services to ensure freedom of action, extend operational reach, and prolong endurance (FM 3-0). EW capabilities that enable the sustainment warfighting function include—

- Protection of sustainment forces from friendly and adversary use of EW in static or mobile environments.
- Enhanced electromagnetic environment situational awareness through the interception, detection, identification, and location of adversary electromagnetic emissions and by providing indications and warnings. (This information can assist in convoy planning, asset tracking, and targeting of potential threats to sustainment operations.)
- Countering improvised explosive devices to support ground lines of communication (includes counter-radio-controlled improvised-explosive-device systems and countering other threats triggered through the electromagnetic spectrum, such as lasers).
- Spectrum deconfliction and emissions control procedures in support of sustainment command and control.
- Electromagnetic spectrum obscuration, low-observability, and multispectral stealth (These capabilities provide protection during sustainment operations).

Command and Control

2-15. The *command and control warfighting function* is the related tasks and systems that support commanders in exercising authority and direction (FM 3-0). EW capabilities that enable the command and control warfighting function include—

- Protection of friendly critical information systems and command and control nodes, personnel, and facilities from the effects of friendly and adversary EW operations.
- Control of friendly EW systems through—
  - Frequency deconfliction.
  - Asset tracking.
  - Employment execution.
  - Reprogramming of EW systems.
  - Registration of all electromagnetic spectrum emitting devices with the spectrum manager (both prior to deployment and when new systems or devices are added to the deployed force).
- The development of EW command and control tools to enhance required coordination between Army and joint EW operations.
- EW operations integration, coordination, deconfliction, and synchronization through the EW working group (see chapter 3).
- Increased commander situational understanding through improved common operational picture input of electromagnetic spectrum- and EW-related information.
- EW operations monitoring and assessment.

Protection

2-16. The *protection warfighting function* is the related tasks and systems that preserve the force so the commander can apply maximum combat power (FM 3-0). EW capabilities and actions that enable the protection warfighting function include—

- Enhanced electromagnetic spectrum situational awareness through the interception, detection, identification, and location of adversary electromagnetic emissions used to providing indications and warnings of threat emitters and radars.
- Denial, disruption, or destruction of electromagnetic-spectrum-triggered improvised explosive devices and enemy air defense systems.
- Deception of enemy forces.
- Electromagnetic spectrum obscuration, low-observability, and multispectral stealth.
- EW countermeasures for platform survivability (air and ground).
- Area denial capabilities (lethal and nonlethal) against personnel, vehicles, and aircraft.
- Protection of friendly personnel, equipment, and facilities from friendly and enemy electronic attack, including friendly information systems and information. (This includes the coordination and use of both airborne and ground-based electronic attack with higher and adjacent units.)

SUMMARY

2-17. Army EW operations provide the land force commander capabilities to support full spectrum operations (offensive, defensive, and stability or civil support operations). EW supports full spectrum operations by applying EW capabilities to detect, deny, deceive, disrupt, or degrade and destroy enemy combat capability and by controlling and protecting friendly use of the electromagnetic spectrum. These capabilities—when applied across the warfighting functions—enable commanders to address a broad set of electromagnetic-spectrum-related targets to gain and maintain an advantage within the electromagnetic spectrum.

Chapter 3
Electronic Warfare Organization

A flexible organizational framework and capable, proficient electronic warfare personnel enable the commander's electronic warfare capability on the battlefield. This chapter discusses a framework that ensures coordination, synchronization, and integration of electronic warfare into full spectrum operations. This electronic warfare organizational framework supports current operations and is adaptable for future operations.

ORGANIZING ELECTRONIC WARFARE OPERATIONS

3-1. Operational challenges across the electromagnetic spectrum are expanding rapidly. As Army electronic warfare (EW) capabilities expand to meet these challenges, the organizational design required to coordinate, synchronize, integrate, and deconflict these capabilities must transform as rapidly. To meet current and future requirements, command and control of EW operations is built around the concept of EW working groups. Figure 3-1, page 3-2, illustrates the EW coordination organizational framework.

ARMY SERVICE COMPONENT COMMAND, CORPS, AND DIVISION LEVELS

3-2. A *working group* is a temporary grouping of predetermined staff representatives who meet to coordinate and provide recommendations for a particular purpose or function (FMI 5-0.1). The EW working group, when established, is responsible to the G-3 through the fires cell. An EW working group usually includes representation from the G-2, G-3, G-5, G-6, and G-7. (Joint doctrine calls this organization the EW coordination cell.) The EW working groups depicted in figure 3-1 (page 3-2) facilitate the internal (Army) and external (joint) integration, synchronization, and deconfliction of EW actions with fires, command and control, movement and maneuver, intelligence, sustainment and protection warfighting functions. Normally, EW working groups do not add additional structure to an existing organization. As depicted in figure 3-1, working groups vary in size and composition based on echelon.

3-3. Normally, the senior EW officer heads the EW working group and is accountable to the G-3 for integrating EW requirements. Working within the fires cell, the EW officer coordinates directly with the fire support coordinator for the integration of EW into the targeting process. This ensures EW capabilities are fully integrated with all other effects. Additional staff representation within EW working groups may include a fire support coordinator, a spectrum manager, a space operations officer, and liaison officers as required. Depending on the echelon, liaisons could include joint, interagency, and multinational representatives. When an Army headquarters serves as the headquarters of a joint task force or joint force land component command, the Army headquarters' working group becomes the joint force EW coordination cell.

3-4. When Army forces are employed as part of a joint or multinational force, they normally have EW representatives supporting higher headquarters' EW coordination organizations. These organizations may include the joint force commander's EW staff or the information operations cell within a joint task force. Sometimes a component EW organization may be designated as the joint EW coordination cell. (Chapter 6 discusses joint electronic warfare operations in more detail.) The overall structure of the combatant force and the level of EW to be conducted determine the structure of the joint EW coordination cell. The organization to accomplish the required EW coordination and functions varies by echelon.

Chapter 3

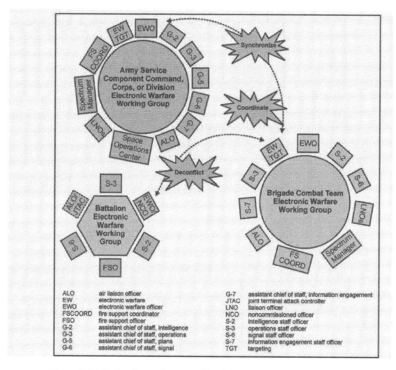

Figure 3-1. Electronic warfare coordination organizational framework

3-5. Regardless of the organizational framework employed, EW working groups perform specific tasks. Table 3-1 (page 3-3) details the functions of the EW working groups by echelon from battalion to Army Service component command. There is no formal organizational framework for EW at the company level (see paragraph 3-9).

Electronic Warfare Organization

Table 3-1. Functions of electronic warfare working groups

EW Working Group	Functions
Division and Above ALO EWO EW targeting G-2 G-3 G-5 G-6 G-7 FSCOORD LNOs Spectrum manager Space support officer	**Peacetime: Division—ASCC** • Conduct long-range electronic warfare planning in support of theater or combatant command requirements. • Integrate electronic warfare into operation plans and concept plans. • Develop electronic warfare supporting plans to operation plans and contingency plans. • Coordinate joint electronic warfare training and exercises. • Develop information and knowledge necessary to support contingency planning (for example, joint restricted frequency list development, spectrum management, and deconfliction). **Wartime: Division—ASCC** • Serve as the joint force land component or joint task force EW working group. • When directed, serve as the jamming control authority. • Develop and promulgate electronic warfare policies and support higher level policies. • Identify and coordinate intelligence support requirements for electronic warfare. • Plan, coordinate, and assess offensive and defensive electronic warfare requirements. • Plan, coordinate, synchronize, deconflict, and assess electronic warfare operations. • Maintain current assessment of electronic warfare resources available to the commander. • Prioritize electronic warfare effects and targets. • Predict effects of friendly and enemy electronic warfare. • Coordinate spectrum management and radio frequency deconfliction with G-6 and J-6. • Plan, assess, and implement friendly electronic security measures. • Plan, coordinate, integrate, and deconflict electronic warfare effects within the operations process.
Brigade S-3 EWO EW targeting FSCOORD S-2 S-6 ALO LNOs S-7 Spectrum manager	**Peacetime:** • Develop electronic warfare supporting requirements to operations plans and exercises. **Wartime:** • Support electronic warfare policies. • Plan, prepare, execute, and assess electronic warfare operations. • Integrate electronic warfare intelligence preparation of the battlefield into the operations process. • Identify and coordinate intelligence support requirements for BCT and subordinate units' electronic warfare operations. • Assess offensive and defensive electronic warfare requirements. • Maintain current assessment of electronic warfare resources available to unit. • Prioritize BCT and subordinate units' electronic warfare targets. • Plan, coordinate, and assess friendly electronic warfare operations. • Implement friendly electronic security measures (for example, electromagnetic spectrum mitigation and network protection). • When directed, serve as the jamming control authority.
Battalion EWO/NCO FSO S-2 S-3 S-6 JTAC	**Peacetime:** • Support BCT electronic warfare requirements to operations and exercises. **Wartime:** • Evaluate electronic warfare offensive, defensive, and support requirements. • Coordinate electronic warfare operations with higher headquarters. • Identify and coordinate intelligence support requirements with higher headquarters. • Execute electronic warfare in support of current operations. • Assess electronic warfare operations.

ALO	air liaison officer	G-7	assistant chief of staff, information engagement staff
ASCC	Army service component command	J-6	communications system directorate of a joint staff
BCT	brigade combat team	JTF	joint task force
EW	electronic warfare	JTAC	joint terminal attack controller
EWO	electronic warfare officer	LNO	liaison officer
FSCOORD	fire support coordinator	NCO	noncommissioned officer
G-2	assistant chief of staff, intelligence	S-2	intelligence staff officer
G-3	assistant chief of staff, operations	S-3	operations staff officer
G-4	assistant chief of staff, logistics	S-6	signal staff officer
G-5	assistant chief of staff, plans	S-7	information engagement staff officer
G-6	assistant chief of staff, signal		

BRIGADE LEVEL

3-6. At the brigade level, the EW officer heads the EW working group and is accountable to the S-3 for integrating EW requirements. Additional staff representation within EW working groups at the brigade combat team level may include the fire support coordinator, EW targeting technician, S-2, S-6, spectrum manager, S-7, and liaison officers as required.

3-7. The EW working group at the brigade combat team coordinates with the higher echelon EW working groups. The brigade working group plays an important role in requesting and integrating joint air and ground EW support. It also manages the brigade's organic EW "fight" within the fires cell. The EW officer works as part of the brigade combat team staff. In this position, the EW officer synchronizes, integrates, and deconflicts brigade combat team EW actions with the EW working group at division level. Although EW falls under the control of the S-3, EW officers are fully immersed in fires targeting and planning to ensure proper use and coordination of EW. See table 3-1, page 3-3, for an outline of the functions of the brigade combat team EW working group.

BATTALION LEVEL

3-8. At the battalion level, the EW officer or noncommissioned officer leads the EW working group and is accountable to the S-3 for integrating EW requirements. Additional staff representation within EW working groups at the battalion level may include the S-2, S-6, fire support officer, and a joint terminal attack controller when assigned. The battalion EW working group coordinates battalion EW operations with the brigade combat team EW working group. See table 3-1, page 3-3, for an outline of the functions of the battalion EW working group.

COMPANY LEVEL

3-9. At the company level, trained EW personnel holding an additional skill identifier of 1K (tactical EW operations) or 1J (operational EW operations) perform several tasks. They advise the commander on the employment of EW equipment, track EW equipment status, assist operators in the use and maintenance of EW equipment, and coordinate with higher headquarters EW working groups.

PLANNING AND COORDINATING ELECTRONIC WARFARE ACTIVITIES

3-10. Key personnel involved in the planning and coordination of EW activities are—
- G-3 and S-3 staff.
- EW officer.
- Fire support coordinator.
- G-2 and S-2 staff.
- G-6 and S-6 staff.
- Electromagnetic spectrum manager.
- Liaisons.

G-3 OR S-3 STAFF

3-11. The G-3 or S-3 staff is responsible for the overall planning, coordination, and supervision of EW activities, except for intelligence. The EW officer is part of the G-3 or S-3 staff. The G-3 or S-3 staff—
- Plans for and incorporates EW into operation plans and orders, in particular within the fire support plan and the information operations plan (in joint operations).
- Tasks EW actions to assigned and attached units.
- Exercises control over electronic attack, including integration of electromagnetic deception plans.

Electronic Warfare Organization

- Directs electronic protection measures the unit will take based on recommendations from the G-6 or S-6, the EW officer, and the EW working group.
- Coordinates and synchronizes EW training with other unit training requirements.
- Coordinates and synchronizes EW training with other unit training requirements.
- Issues EW support tasks within the unit intelligence, surveillance, and reconnaissance plan. These tasks are according to the collection plan and the intelligence synchronization matrices developed by the G-2 or S-2 and the collection manager.
- Coordinates with the EW working group to ensure planned EW operations support the overall tactical plan.
- Integrates electronic attack as a form of fires within the fires cell.

ELECTRONIC WARFARE OFFICER

3-12. As a member of the G-3 or S-3 staff, the EW officer plans, coordinates, and supports the execution of EW. The EW officer—

- Leads the EW working group.
- Plans, coordinates, and assesses EW offensive, defensive, and support requirements.
- Supports the G-2 or S-2 during intelligence preparation of the battlefield.
- Supports the fire support coordinator to ensure electronic attack fires are integrated with all other effects.
- Plans, assesses, and implements friendly electronics security measures.
- Prioritizes EW effects and targets with the fire support coordinator.
- Plans and coordinates EW operations across functional and integrating cells.
- Deconflicts EW operations with the spectrum manager.
- Maintains a current assessment of available EW resources.
- Participates in other cells and working groups (as required) to ensure EW integration.
- Serves as EW subject matter expert on existing EW rules of engagement.
- When designated, serves as the jamming control authority.
- Prepares, submits for approval, and supervises the issuing and implementation of fragmentary orders for EW operations.

G-2 OR S-2 STAFF

3-13. The G-2 or S-2 staff advises the commander and staff on the intelligence aspects of EW. The G-2 or S-2 staff—

- Provides threat data to support programming of unit EW systems and deconfliction of their use by the EW working group.
- Ensures that electronic order of battle requirements are included in the intelligence collection plan.
- Determines enemy EW organizations, disposition, capabilities, and intentions via collection and analysis.
- Determines enemy EW vulnerabilities and high-value targets.
- Assesses effects of friendly EW operations on the enemy.
- Helps prepare the intelligence-related portion of the EW running estimate.
- Provides input to the restricted frequency list by recommending guarded frequencies.
- Provides updates on the rapid electronic order of battle.
- Maintains appropriate threat EW databases.
- Works with the EW working group to ensure that intelligence collection is synchronized with EW requirements and deconflicted with planned EW actions. Ensures that EW threat data is deconflicted with friendly electromagnetic spectrum needs.

NETWORK OPERATIONS OFFICER

3-14. The network operations officer (in the G-6 or S-6 staff) coordinates the communications network for the following services:
- Preparing the electronic protection policy on behalf of the commander.
- Assisting in preparing EW plans and orders.
- Reporting all enemy electronic attack activity detected by friendly communications and electronics elements to the EW working group for counteraction.
- Assisting the unit EW officer with resolving EW systems maintenance and communications fratricide problems.

SPECTRUM MANAGER

3-15. The spectrum manager coordinates electromagnetic spectrum use for a wide variety of communications and electronic resources. The spectrum manager—
- Issues the signal operating instructions.
- Provides all spectrum resources to the task force.
- Coordinates for spectrum usage with higher echelon G-6 or S-6, and applicable host-nation and international agencies as necessary.
- Coordinates the preparation of the restricted frequency list and issuance of emissions control guidance.
- Coordinates frequency allotment, assignment, and use.
- Coordinates electromagnetic deception plans and operations in which assigned communications resources participate.
- Coordinates measures to reduce electromagnetic interference.
- Coordinates with higher echelon spectrum managers for electromagnetic interference resolution that cannot be resolved internally.
- Assists the EW officer in issuing guidance in the unit (including subordinate elements) regarding deconfliction and resolution of interference problems between EW systems and other friendly systems.
- Participates in the EW working group to deconflict friendly electromagnetic spectrum requirements with planned EW operations and intelligence collection.

SUMMARY

3-16. The organizational framework for EW coordination and functions varies by echelon. The necessity to form an EW working group is largely based on the overall structure of the combatant force and the level of EW to be conducted. During unified actions, other Service EW officers, signals intelligence officers, and EW asset representatives are invaluable to Army EW working groups in the planning, preparation, execution, and assessment of EW operations. As Army EW capabilities and concepts for employment continue to evolve, so do the organizational designs that ensure their effective command and control and execution in support of operations.

Chapter 4
Electronic Warfare and the Operations Process

The *operations process* consists of the major command and control activities performed during operations: planning, preparing, executing, and continuously assessing the operation. The commander drives the operations process (FM 3-0). These activities occur continuously throughout an operation, overlapping and recurring as required (see figure 4-1). The staff electronic warfare officer is actively involved in the operations process. Electronic warfare planning, preparation, execution, and assessment require collective expertise from operations, intelligence, signal, and battle command. The electronic warfare officer—through the unit's electronic warfare working group—integrates efforts across the warfighting functions. This ensures that electronic warfare operations support the commander's objectives.

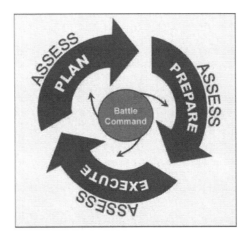

Figure 4-1. The operations process

SECTION I — ELECTRONIC WARFARE PLANNING

4-1. Electronic warfare (EW) planning is based on three main considerations. The first is applying the military decisionmaking process (MDMP). EW planners understand and follow its seven steps. In a time-constrained environment they still follow all seven steps, abbreviating the MDMP process appropriately. Additionally, EW planners apply EW integrating processes. They understand how EW actions contribute to operations. They integrate and synchronize EW activities starting with planning and continuing throughout operations. Finally, EW planners apply EW employment considerations according to the characteristics of EW capabilities.

Chapter 4

THE MILITARY DECISIONMAKING PROCESS

4-2. EW planning minimizes fratricide and optimizes operational effectiveness during execution. Therefore, EW planning occurs concurrently with other operational planning during the MDMP. The MDMP synchronizes several processes, including intelligence preparation of the battlefield (IBP) (see FM 34-130), the targeting process (see FM 6-20-10), and risk management (see FM 5-19). These processes occur continuously during operations.

4-3. Depending on the organizational echelon, the staff EW officer leads EW planning through the EW working group. (The EW working group at echelons above brigade is sometimes referred to as an EW coordination cell.) An EW working group is normally supported by representatives from the G-2 or S-2, G-3 or S-3, G-6 or S-6, and other staff as required. Other staff representatives can include the fire support coordinator or fire support officer, spectrum manager, air liaison officer, space officer, and liaison officers. Paragraphs 4-5 through 4-33 outline key EW contributions to the processes and planning actions that occur during the seven steps of the MDMP. (FM 5-0 discusses the MDMP.)

RECEIPT OF MISSION

4-4. Commanders begin the MDMP upon receiving or anticipating a new mission. During this first step, commanders issue their initial guidance and initial information requirements or commander's critical information requirements.

4-5. Upon receipt of a mission, the staff EW officer alerts the staff members supporting the EW working group. The EW officer and support staff begin to gather the resources required for mission analysis. Resources might include a higher headquarters operation order or plan, maps of the area of operations, electronic databases, required field manuals and standing operating procedures, current running estimates, and reachback resources (see appendix F). The EW officer also provides input to the staff's initial assessment and updates the EW running estimate. As part of this update, the EW officer identifies all friendly EW assets and resources and their status. The EW officer also provides this information throughout the operations process. This includes monitoring, tracking, and seeking out information relating to EW operations to assist the commander and staff.

MISSION ANALYSIS

4-6. Planning includes a thorough mission analysis. Both the process and products of mission analysis help commanders refine their situational understanding and determine their restated mission. (See FM 5-0 for more details.) The EW officer and supporting members of the EW working group contribute to the overall mission analysis by participating in IPB and through the planning actions discussed in paragraphs 4-7 through 4-14. (Paragraphs 4-35 to 4-40 discuss EW input to IPB during operations.)

4-7. The EW officer and EW working group members—
- Convene the appropriate EW working group.
- Determine known facts, status, or conditions of forces capable of EW operations as defined in the commander's planning documents, such as a warning order or operation order.
- Identify EW planning support requirements and develop support requests as needed.

4-8. The EW officer and EW working group members support the G-2 and S-2 in IPB by—
- Determining the threat's dependence on the electromagnetic spectrum.
- Determining the threat's EW capability.

Electronic Warfare and the Operations Process

- Determining the threat's intelligence system collection capability.
- Determining which threat vulnerabilities relate to the electromagnetic spectrum.
- Determining how the operational environment affects EW operations using the operational variables and mission variables as appropriate.
- Initiating, refining, and validating information requirements and requests for information.

4-9. The EW officer and EW working group members—
- Determine facts and develop necessary assumptions relevant to EW such as the status of EW capability at probable execution and time available.
- Analyze the commander's mission and intent from an EW perspective.
- Identify constraints relevant to EW—
  - Actions EW operations must perform.
  - Actions EW operations cannot perform.
  - Other constraints.
- Analyze mission, enemy, terrain and weather, troops and support available, time available and civil considerations from the EW perspective.

4-10. The EW officer and EW working group members determine enemy and friendly centers of gravity and list their critical capabilities, requirements, and vulnerabilities from an EW perspective. (They determine how EW capabilities can best attack an enemy's command and control system.) The center of gravity analysis process outlined in figure 4-2 helps identify and list the critical vulnerabilities of enemy centers of gravity. The EW officer and EW working group members also list the critical requirements associated with the identified command and control critical capability (or command and control nodes) and then identify the critical vulnerabilities associated with the critical requirements. Through this process, the EW officer and EW working group members help determine which vulnerabilities can be engaged by EW capabilities to produce a decisive outcome.

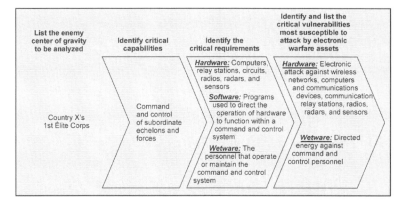

Figure 4-2. Example of analysis for an enemy center of gravity

4-11. Additionally, the EW officer and EW working group members determine how EW can help protect friendly centers of gravity. The center of gravity analysis process as outlined in figure 4-2 can also be used help identify critical vulnerabilities of friendly centers of gravity. The EW officer and EW working group members list the critical requirements associated with the identified friendly command and control critical capability. Then, the EW officer and EW working group members identify the critical vulnerabilities associated with the critical requirements. These vulnerabilities can help determine how to best use EW capabilities to defend or protect friendly centers of gravity from enemy attack. Key to this portion of the analysis is to assess the potential impact of EW operations on friendly information systems such as electromagnetic interference.

4-12. The EW officer and EW working group members identify and list—
- High-value targets that can be engaged by EW capabilities.
- Tasks that EW forces perform according to EW subdivision (electronic attack, electronic warfare support, and electronic protection) in support of the warfighting functions. These include—
  - Determining specified EW tasks.
  - Determining implied EW tasks.

4-13. The EW officer and EW working group members—
- Conduct initial EW force structure analysis to determine if sufficient assets are available to perform the identified EW tasks. (If organic assets are insufficient, they draft requests for support and augmentation.)
- Conduct an initial EW risk assessment and review the risk assessment done by the entire working group.
- Provide EW perspective in the development of the commander's restated mission.
- Assist in development of the mission analysis briefing for the commander.

4-14. By the conclusion of mission analysis, the EW officer and EW working group members generate or gather the following products and information:
- The initial information requirements for EW operations.
- A rudimentary command and control nodal analysis of the enemy.
- The list of EW tasks required to support the mission.
- A list of assumptions and constraints related to EW operations.
- The planning guidance for EW operations.
- EW personnel augmentation or support requirements.
- An update of the EW running estimate.
- EW portion or input to the commander's restated mission.

COURSE OF ACTION DEVELOPMENT

4-15. After receiving the restated mission, commander's intent, and commander's planning guidance, the staff develops courses of action (COAs) for the commander's approval. Figure 4-3 depicts the required input to COA development and identifies the key contributions made by the EW officer and EW working group members during the process and output stages (center and right of figure 4-3). The actions the EW officer and EW working group members perform to support COA development are discussed in more detail in paragraphs 4-16 through 4-20.

Electronic Warfare and the Operations Process

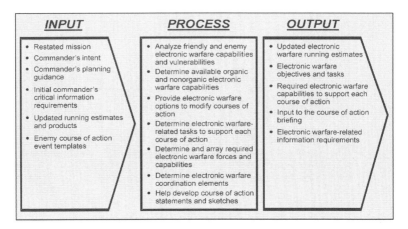

Figure 4-3. Course of action development

4-16. The EW officer and EW working group members contribute to COA development through the following planning actions—
- Determining which friendly EW capabilities are available to support the operation, including organic and nonorganic capabilities for planning.
- Determining possible friendly and enemy EW operations, including identifying friendly and enemy vulnerabilities.

4-17. Additionally, the EW officer and EW working group members help develop initial COA options by—
- Identifying COA options that may be feasible based on their functional expertise (while brainstorming of COAs).
- Providing options to modify a COA to enable accomplishing a requirement within the EW area of expertise.
- Identifying information (relating to EW options) that may impact other functional areas and sharing that information immediately.
- Identifying the EW-related tasks required to support the COA options.

4-18. The EW officer and EW working group members determine the forces required for mission accomplishment by—
- Determining the EW tasks that support each COA and how to perform those tasks based on available forces and capabilities. (Available special technical operations capabilities are considered in this analysis.)
- Providing input and support to proposed deception options.
- Ensuring the EW options provided in support of all possible COAs meet the established screening criteria.

4-19. The EW officer and EW working group members identify EW supporting tasks and their purpose in supporting any decisive, shaping, and sustaining operations as each COA is developed. These EW tasks include those—
- Focused on defeating the enemy.
- Required to protect friendly force operations.

4-20. The EW officer and EW working group members assist in developing the COA briefing as required. By the conclusion of COA development, the EW officer and EW working group members generate or gather the following products and information:
- A list of EW objectives and desired effects related to the EW tasks.
- A list of EW capabilities required to perform the stated EW tasks for each COA.
- The information and intelligence requirements for performing the EW tasks in support of each COA.
- An update to the EW running estimate.

COURSE OF ACTION ANALYSIS (WAR-GAMING)

4-21. The COA analysis allows the staff to synchronize the elements of combat power for each COA and to identify the COA that best accomplishes the mission. It helps the commander and staff to—
- Determine how to maximize the effects of combat power while protecting friendly forces and minimizing collateral damage.
- Further develop a visualization of the battle.
- Anticipate battlefield events.
- Determine conditions and resources required for success.
- Determine when and where to apply force capabilities.
- Focus IPB on enemy strengths and weaknesses as well as the desired end state.
- Identify coordination needed to produce synchronized results.
- Determine the most flexible COA.

Paragraphs 4-22 to 4-23 discuss actions the EW officer and EW working group members perform to support COA analysis. (See FM 5-0 for more information on war-gaming.)

4-22. During COA analysis, the EW officer and EW working group members synchronize EW actions and assist the staff in integrating EW capabilities into each COA. The EW officer and EW working group members address how each EW capability supports each COA. They apply these capabilities to associated time lines, critical events, and decision points in the synchronization matrix (see table 4-1). During this planning phase, the EW officer and EW working group members aim to—
- Analyze each COA from an EW functional perspective.
- Recommend any EW task organization adjustments.
- Identify key EW decision points.
- Provide EW data for synchronization matrix.
- Recommend EW priority intelligence requirements.
- Identify EW supporting tasks to any branches and sequels.
- Identify potential EW high-value targets.
- Assess EW risks created by telegraphing intentions, allowing time for enemy to mitigate effects, unintended effects of electronic attack, and the impact of asset or capability shortfalls.

4-23. By the conclusion of COA analysis (war-gaming), the EW officer and EW working group members generate or gather the following products and information:
- The EW data for the synchronization matrix.
- The EW portion of the branches and sequels.
- A list of high-value targets related to EW.
- A list of commander's critical information requirements related to EW.
- The risk assessment for EW operations in support of each COA.
- An update to the EW running estimate.

Electronic Warfare and the Operations Process

Table 4-1. Sample input to synchronization matrix

	TIME/EVENT	H - 8	H - hour	H + 8
	Enemy Actions	Enemy monitors movements	Defends from Security Zone	Commits reserve
	Decision Points	Launch deep attack		
M A N E U V E R	1st Brigade	Move on route Paula	Cross line of departure	Seize objective Nick
	2d Brigade	Move on route Mike	Cross line of departure	Seize objective Dave
	3rd Brigade	Move on route Sean		Forward passage of lines with 1st Brigade
	Aviation Brigade	Deep attack on objective Rose at H - 1		
	Division Cavalry		Screen northern flank	
	Fires Brigade	Preparation fires initiated at H - 5		
F I R E S	Air Defense	Weapons hold	Weapons tight	Weapons tight
	C2W - EA - CNA - Physical Attack - CNE - ES	- Locate enemy ISR on maneuver routes - Deny and disrupt enemy ISR of maneuver routes at H - .5 to H - hour - Disrupt and destroy known enemy C2 nodes and IADS	- Activate CREW systems - Jamming (to disrupt/deny enemy C2 nodes) - Electronic deception - Provide indications and warnings to maneuver brigades	Disrupt and destroy enemy C2 system
C2			Tactical CP with lead brigade	

C2	command and control	EA	electronic attack
C2W	command and control warfare	ES	electronic warfare support
CNA	computer network attack	EW	electronic warfare
CNE	computer network exploitation	H-hour	specific time an operation or exercise begins
CP	command post	IADS	integrated air defense system
CREW	counter radio-controlled improvised explosive device electronic warfare	ISR	intelligence, surveillance, and reconnaissance

Note: This is not complete. Its intent is to show how EW can be integrated into a synchronization matrix.

Chapter 4

COURSE OF ACTION COMPARISON

4-24. COA comparison starts with all staff members analyzing and evaluating the advantages and disadvantages of each COA from their perspectives. Staff members present their findings for the others' consideration. Using the evaluation criteria developed during COA analysis, the staff outlines each COA, highlighting its advantages and disadvantages. Comparing the strengths and weaknesses of the COAs identifies their advantages and disadvantages with respect to each other. (See FM 5-0 for further discussion of COA comparison).

4-25. During COA comparison, the EW officer and EW working group members compare COAs based on the EW-related advantages and disadvantages (see center of figure 4-4). Typically, planners use a matrix to assist in the COA comparisons. The EW officer may develop an EW functional matrix to compare the COAs or to use the decision matrix developed by the staff. Regardless of the matrix used, the evaluation criteria developed prior to war-gaming are used to compare the COAs. Normally, the chief of staff or executive officer weights each criterion used for the evaluation based on its relative importance and the commander's guidance. (See FM 5-0 for more information on COA comparison and a sample decision matrix.)

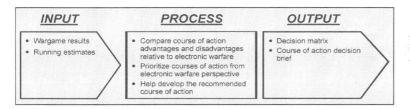

Figure 4-4. Course of action comparison

4-26. By the conclusion of COA comparison, the EW officer and EW working group members generate or gather the following products and information:
- A list of the pros and cons for each COA relative to EW.
- A prioritized list of the COAs from an EW perspective.
- An update to the EW running estimate if required.

COURSE OF ACTION APPROVAL

4-27. The COA approval process has three components. First, the staff recommends a COA, usually in a decision briefing. Second, the commander decides which COA to approve. Lastly, the commander issues the final planning guidance.

4-28. During COA approval, the EW officer supports the development of the COA decision briefing and the development of the warning order as required. If possible, the EW officer attends the COA decision briefing to receive the commander's final planning guidance. If unable to attend the briefing, the EW officer receives the final planning guidance from the G-3 or S-3. The final planning guidance is critical in that it normally provides—
- Refined commander's intent.
- New commander's critical information requirements to support the execution of the chosen COA.
- Risk acceptance.
- Guidance on priorities for the elements of combat power, orders preparation, rehearsal, and preparation.

4-29. After the COA decision has been made, the EW officer and EW working group members generate or gather the following products and information:
- An updated command and control nodal analysis of the enemy relevant to the selected COA.
- Required requests for information to refine the enemy command and control nodal architecture.
- Latest electronic order of battle tailored to the selected COA.
- Any new direction provided in the refined commander's intent.
- A list of any new commander's critical information requirements that can be used in support of EW operations.
- The warning order to assist developing EW operations required to support the operation order or plan.
- Refined input to the initial intelligence, surveillance, and reconnaissance (ISR) plan, including—
  - Any additional specific EW information requirements.
  - Updated potential collection assets for the unit's ISR plan.

ORDERS PRODUCTION

4-30. Orders production consists of the staff preparing the operation order or plan by converting the selected COA into a clear, concise concept of operations. The staff also provides supporting information that enables subordinates to execute and implement risk controls. They do this by coordinating and integrating risk controls into the appropriate paragraphs and graphics of the order.

4-31. During orders production, the EW officer provides the EW operations input for several sections of the operation order or plan. See appendix B for the primary areas for EW operations input within an Army order or plan. The primary areas for EW input in a joint order, if required, also are shown in appendix B. (See CJCSM 3122.03C for the Joint Operation Planning and Execution System format).

DECISIONMAKING IN A TIME-CONSTRAINED ENVIRONMENT

4-32. In a time-constrained environment, the staff might not be able to conduct a detailed MDMP. The staff may choose to abbreviate the process as described in FM 5-0. The abbreviated process uses all seven steps of the MDMP in a shortened and less detailed manner.

4-33. The EW officer and core members of the EW working group meet as a regular part of the unit battle rhythm. However, the EW officer calls unscheduled meetings if situations arise that require time-sensitive planning. Regardless of how much they abbreviate the planning process, the EW officer and supporting members of the EW working group always—
- Update the EW running estimate in terms of assets and capabilities available.
- Update essential EW tasks with the requirements of the commander's intent.
- Coordinate support requests and intelligence requirements with appropriate staff elements and outside agencies.
- Provide EW input to fragmentary orders through the G-3 or S-3 as necessary to drive timely and effective EW operations.
- Deconflict planned EW actions with other uses of the spectrum, such as communications.
- Synchronize electronic attack and EW support actions.
- Synchronize other intelligence collection in support of EW requirements.
- Deconflict EW activities specifically with aviation operations.
- Synchronize EW support to the command and control warfare and information protection information tasks.

Chapter 4

THE INTEGRATING PROCESSES AND CONTINUING ACTIVITIES

4-34. Commanders use several integrating processes and continuing activities to synchronize operations throughout the operations process. (See figure 4-5.) The EW officer ensures EW operations are fully synchronized and integrated within these processes and continuing activities. Other staff members supporting the EW working group assist the EW officer. Paragraphs 4-35 through 4-52 outline some key integrating processes and continuing activities. These processes and activities require EW officer involvement throughout the operations process.

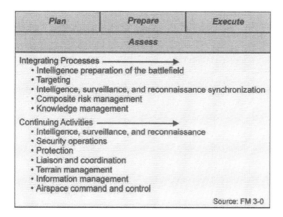

Figure 4-5. Integrating processes and continuing activities

INTELLIGENCE PREPARATION OF THE BATTLEFIELD

4-35. *Intelligence preparation of the battlefield* is the systematic, continuous process of analyzing the threat and environment in a specific geographic area. Intelligence preparation of the battlefield is designed to support the staff estimate and military decisionmaking process. Most intelligence requirements are generated as a result of the intelligence preparation of the battlefield process and its interrelation with the decisionmaking process (FM 34-130). The G-2 or S-2 leads IPB planning with participation by the entire staff. This planning activity is used to define and understand the operational environment and the options it presents to friendly and adversary forces. Only one IPB planning activity exists within each headquarters; all affected staff cells participate. (FM 2-0 provides more information on IPB.) Paragraphs 4-36 through 4-40 discuss how the EW officer and the EW working group support IPB during operations.

4-36. In addition to the input provided to the initial IPB (during step 2 of mission analysis), the EW officer supports IPB throughout the operations process by providing input related to EW operations. (See figure 4-6.) This input includes (but is not limited to) the following EW considerations:
- Evaluating the operational environment from an EW perspective.
- Describing how the effects of the operational environment may impact EW operations.
- Evaluating the threat's capabilities; doctrinal principles; and tactics, techniques, and procedures from an EW perspective.
- Determining threat COAs.

Electronic Warfare and the Operations Process

4-37. When evaluating the operational environment from an EW perspective, the EW officer—
- Determines the electromagnetic environment within the defined physical environment:
  - Area of operations.
  - Area of influence.
  - Area of interest.
- Uses electronic databases to identify gaps.
- Identifies adversary fixed EW sites such as EW support and electronic attack sites.
- Identifies airfields and installations that support, operate, or house adversary EW capabilities.
- In coordination with the G-2 or S-2 and G-6 or S-6, helps identify enemy electromagnetic spectrum usage and requirements within the area of operations and area of interest.

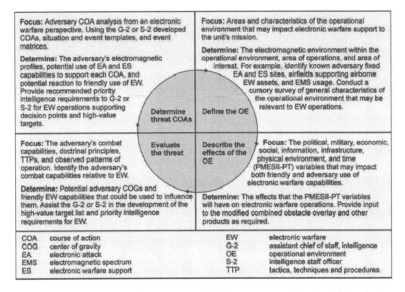

Figure 4-6. Electronic warfare support to intelligence preparation of the battlefield

4-38. When describing how the variables of the operational environment may impact EW operations, the EW officer—
- Focuses on characteristics of both the land and air domains using the factors of observation and fields of fire, avenues of approach, key and decisive terrain, obstacles, and cover and concealment.
- Identifies key terrain that may provide protection for communications and target acquisition systems from exploitation or disruption.
- Identifies how terrain affects line of sight, including effects on both communications and non-communications emitters.
- Evaluates how vegetation affects radio wave absorption and antenna height requirements.
- Locates power lines and their potential to interfere with radio waves.
- Assesses most likely and most dangerous avenues of approach (air, ground) and where EW operations would likely be positioned to support these approaches.

Chapter 4

- If operating within urban terrain, considers how the infrastructure—power plants, power grids, structural heights, and communications and media nodes—may restrict or limit EW capabilities.
- Assists the G-2 or S-2 with the development of a modified combined obstacle overlay.
- Determines how weather—visibility, cloud cover, rain, and wind—may affect ground-based and airborne EW operations and capabilities (for example, no-go weather conditions at an airborne EW launch and recovery base).
- Considers all other relevant aspects of the operational environment that affect EW operations, using the operational variables (PMESII-PT—political, military, economic, social, information, infrastructure, physical environment, and time) and mission variables (METT-TC—mission, enemy, terrain and weather, troops and support available, time available, and civil considerations).

4-39. When evaluating enemy capabilities, the EW officer and supporting staff examine doctrinal principles; tactics, techniques and procedures; and observed patterns of operation from an EW perspective. The EW officer—

- Uses the operational variables (PMESII-PT) and mission variables (METT-TC) to help determine the adversary's critical nodes.
- Collects the required data—operational net assessments, electronic order of battle, and electronic databases—to template the command and control critical nodes and the systems required to support and maintain them.
- Assists the G-2 in determining the adversary's EW-related threat characteristics (order of battle) by identifying—
  - Types of communications equipment available.
  - Types of noncommunications emitters.
  - Surveillance and target acquisition assets.
  - Technological sophistication of the threat.
  - Communications network structure.
  - Frequency allocation techniques.
  - Operation schedules.
  - Station identification methods.
  - Measurable characteristics of communications and noncommunications equipment.
  - Command, control, and communications structure of the threat.
  - Tactics from a communication perspective. Examples are how the enemy deploys command, control, and communications assets; whether or not communications systems are remote; and the level of discipline in procedures, communications security, and operations security.
  - Electronic deception capabilities.
  - Reliance on active or passive surveillance systems
  - Electromagnetic profiles of each node.
  - Unique electromagnetic spectrum signatures.
- Assists the G-2 or S-2 in center of gravity analysis. Helps identify the critical system nodes of the center of gravity and determines what aspects of the system should be engaged, exploited, or attacked to modify the system's behavior or to achieve a desired effect.
- Identifies organic and nonorganic EW capabilities available to achieve desired effects on identified high-value targets.
- Submits initial EW-related requests for information that describe the intelligence support required to support EW operations.
- Obtains the high-value target list, threat templates, and initial priority intelligence requirements list to assist in follow-on EW planning.

Electronic Warfare and the Operations Process

4-40. When determining adversary COAs, the EW officer—
- Assists the G-2 or S-2 in development of adversary COAs.
- Provides EW input to the situation templates.
- Ensures event templates include EW named areas of interests.
- Assists in providing EW options for target areas of interest.
- Assists in providing EW options to support decision points.
- Provides EW input to the event template and event matrix.

TARGETING

4-41. *Targeting* is the process of selecting and prioritizing targets and matching the appropriate response to them, considering operational requirements and capabilities (JP 3-0). A decide, detect, deliver, and assess methodology is used to direct friendly forces to attack the right target with the right asset at the right time. (See figure 4-7.) Targeting provides an effective method to match the friendly force capabilities against targets. Commander's intent plays a critical role in the targeting process. The targeting working group strives to thoroughly understand the commander's intent to ensure the commander's intended effects on targets are achieved.

4-42. An important part of targeting is identifying potential fratricide situations and performing the coordination measures to manage and control the targeting effort positively. The targeting working group and staff incorporate these measures into the coordinating instructions and appropriate annexes of the operation plans and orders. (FM 6-20-10 has more information on targeting.)

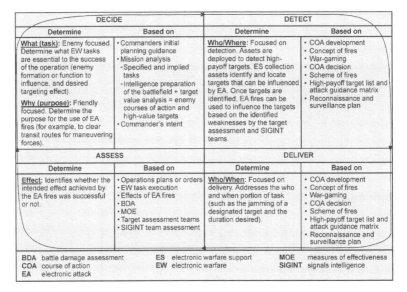

Figure 4-7. Electronic warfare in the targeting process

Chapter 4

4-43. The EW officer thoroughly integrates electronic attack in the targeting process and integrates electronic attack fires into all appropriate portions of the operation plan, operation order, and other planning products. In support of EW targeting, the EW officer—

- Helps the targeting working group determine electronic attack requirements against specific high-payoff targets and high-value targets.
- Ensures electronic attack can meet the desired effect (in terms of the targeting objective).
- Coordinates with the signals intelligence staff element through the collection manager to satisfy EW support and electronic attack information requirements.
- Prepares the EW tab and the EW portion of the command and control warfare tab to the fires appendix.
- Provides electronic attack mission management through the tactical operations center or joint operations center and the tactical air control party (for airborne electronic attack).
- Provides electronic attack mission management as the jamming control authority for ground or airborne electronic attack when designated.
- Prepares and coordinates the EW annex for operation plans and operation orders.
- Determines and requests theater Army electronic attack support.
- Recommends to the G-3 or S-3 and the fire support coordinator or fire support officer whether to engage a target with electronic attack.
- Expedites electromagnetic interference reports to the targeting working group. (See appendix D for information on electromagnetic interference reporting.)

Decide

4-44. Decide is the first step in the targeting process. This step provides the overall focus for fires, a targeting plan, and some of the priorities for intelligence collection. As part of the staff in the fires cell, the EW officer assists the targeting working group in planning the target priorities for each phase and critical events of the operation. Initially, the targeting working group does not develop electronic attack targets using any special technique or separately from targets for physical destruction. However, as the process continues, these targets are passed through intelligence organizations and further planned using ISR procedures. The planned use of electronic attack is integrated into the standard targeting products (graphic or text-based). Products that involve electronic attack planning may include—

- High-payoff target list.
- Attack guidance matrix.
- Appendix 4 (Electronic Warfare) to Annex P (Information Operations) of the operation order. (At the time this manual was written, this was the current doctrine for operation orders. This appendix will be revised upon publication of the revised FM 5-0.)

Detect

4-45. Based on what the targeting working group identified as high-payoff targets during the decide step, collection assets are then deployed to detect them. The intelligence enterprise pairs assets to targets based on the collection plan and the current threat situation. When conducting electronic attack operations in support of command and control warfare, ISR units perform EW support tasks linked to and working closely with the electronic attack missions. Electronic warfare support units (with support from the target assessment and signals intelligence staff elements) provide the data—location, signal strength, and frequency of the target—to focus electronic attack assets on the intended target. These assets also identify the command and control system vulnerabilities open to attack by electronic attack assets.

Deliver

4-46. Once friendly force capabilities identify, locate, and track the high-payoff targets, the next step in the process is to deliver fires against those targets. Electronic attack assets must satisfy the attack guidance developed during the decide step. Close coordination between those conducting EW support and electronic attack is critical during the engagement. The EW officer facilitates this coordination and ensures electronic attack fires are fully synchronized and deconflicted with other fires. The EW officer remains aware of the potential for unintended effects between adjacent units when conducting electronic attack. The EW officer continually coordinates with adjacent unit EW officers to mitigate and deconflict these effects during cross-boundary operations. Normally, the G-3, S-3, or fire support coordinator provides requirements and guidance for this coordination and synchronization in the attack guidance matrix, intelligence synchronization matrix, spectrum management plan, and the EW input to the operation plan or operation order annexes and appendixes.

Assess

4-47. Once the target as been engaged, the next step is to assess the engagement's effectiveness. This is done through combat assessment, which involves determining the effectiveness of force employment during military operations. It consists of three elements:
- Munitions effects assessment.
- Battle damage assessment.
- Re-attack recommendations.

4-48. The first two elements, munitions effects assessment and battle damage assessment, are used to inform the commander on the effects achieved against targets and target sets. From this information, the G-2 or S-2 continues to analyze the threat's ability to further conduct and sustain combat operations (sometimes articulated in terms of the effects achieved against the threat's centers of gravity). The last element involves the assessment and recommendation whether or not to re-attack the targets.

4-49. The assessment of a jamming mission used against an enemy's command and control system is unlike fires that can be observed visually. The signals intelligence staff element and units executing the electronic attack mission coordinate continuously to assess mission effectiveness. Close coordination between sensor and shooter allows instant feedback on the success or failure of the intended jamming effects. It also can quickly provide the necessary adjustments to produce desired effects.

INTELLIGENCE, SURVEILLANCE, AND RECONNAISSANCE SYNCHRONIZATION

4-50. *Intelligence, surveillance, and reconnaissance synchronization* is the task that accomplishes the following: analyzes information requirements and intelligence gaps; evaluates available assets internal and external to the organization; determines gaps in the use of those assets; recommends intelligence, surveillance, and reconnaissance assets controlled by the organization to collect on the commander's critical information requirements; and submits requests for information for adjacent and higher collection support (FM 3-0). ISR synchronization considers all assets—both internal and external to the organization. It identifies information gaps and the most appropriate assets for collecting information to fill them.

4-51. Planning for ISR operations begins during mission analysis. Although led by the G-3 or S-3, it is supported by the entire staff, subordinate units, and external partners. ISR operations collect, process, store, display, and disseminate information from a multitude of collection sources. The staff thoroughly understands, integrates, and synchronizes the ISR plan across all echelons.

4-52. The EW officer ensures the ISR plan supports the EW-related information requirements determined during the planning process. The EW officer coordinates these requirements with the signals intelligence staff element through the G-2 or S-2.

EMPLOYMENT CONSIDERATIONS

4-53. EW has specific ground-based, airborne, and functional (electronic attack, electronic warfare support, or electronic protection) employment considerations. The EW officer ensures EW-related employment considerations are properly articulated early in the operations process. Each capability employed has certain advantages and disadvantages. The staff plans for all of these before executing EW operations.

Chapter 4

GROUND-BASED ELECTRONIC WARFARE CONSIDERATIONS

4-54. Ground-based EW capabilities support the commander's scheme of maneuver. Ground-based EW equipment can be employed by a dismounted Soldier or on highly mobile platforms. Due to the short-range nature of tactical signals direction finding, electronic attack assets are normally located in the forward areas of the battlefield, with or near forward units.

4-55. Ground-based EW capabilities have certain advantages. They provide direct support to maneuver units (for example, through counter-radio-controlled improvised-explosive-device EW and communications or sensor jamming). Ground-based EW capabilities support continuous operations and respond quickly to EW requirements of the ground commander. However, to maximize the effectiveness of ground-based EW capabilities, maneuver units must protect EW assets from enemy ground and aviation threats. EW equipment should be as survivable and mobile as the force it supports. Maneuver units must logistically support the EW assets, and supported commanders must clearly identify EW requirements.

4-56. Ground-based EW capabilities have certain limitations. They are vulnerable to enemy attack and can be masked by terrain. They are vulnerable to enemy electromagnetic deceptive measures and electronic protection actions. In addition, they have distance or propagation limitations against enemy electronic systems.

AIRBORNE ELECTRONIC WARFARE CONSIDERATIONS

4-57. While ground-based and airborne EW planning and execution are similar, they significantly differ in their EW employment time. Airborne EW operations are conducted at much higher speeds and generally have a shorter duration than ground-based operations. Therefore, the timing of airborne EW support requires detailed planning.

4-58. Airborne EW requires the following:
- A clear understanding of the supported commander's EW objectives.
- Detailed planning and integration.
- Ground support facilities.
- Liaisons between the aircrews of the aircraft providing the EW support and the aircrews or ground forces being supported.
- Protection from enemy aircraft and air defense systems.

4-59. Airborne EW capabilities have certain advantages. They can provide direct support to other tactical aviation missions such as suppression of enemy air defenses, destruction of enemy air defenses, and employment of high-speed antiradiation missiles. They can provide extended range over ground-based assets. Airborne EW capabilities can provide greater mobility and flexibility than ground-based assets. In addition, they can support ground-based units in beyond line-of-sight operations.

4-60. The limitations associated with airborne EW capabilities are time-on-station considerations, vulnerability to enemy electronic protection actions, electromagnetic deception techniques, and limited assets (support from nonorganic EW platforms need to be requested).

ELECTRONIC ATTACK CONSIDERATIONS

4-61. Electronic attack includes both offensive and defensive activities. (Chapter 1 provides a full definition of electronic attack). These activities differ in their purpose. Defensive electronic attack protects friendly personnel and equipment or platforms. Offensive electronic attack denies, disrupts, or destroys enemy capability. In either case, certain considerations are involved in planning for employing electronic attack:
- Friendly communications.
- Intelligence collection.
- Other effects.
- Nonhostile local electromagnetic spectrum use.

- Hostile intelligence collection.
- Persistency of effect.

4-62. The EW officer, the G-2 or S-2, the G-3 or S-3, the G-6 or S-6, the spectrum manager, and the G-7 or S-7 coordinate closely to avoid friendly communications interference that can occur when using EW systems on the battlefield. Coordination ensures that electronic attack systems frequencies are properly deconflicted with friendly communications and intelligence systems or that ground maneuver and friendly information tasks are modified accordingly.

4-63. The number of information systems, EW systems, and sensors operating simultaneously on the battlefield makes deconfliction with communications systems a challenge. The EW officer, the G-2 or S-2, the G-6 or S-6, and the spectrum manager plan and rehearse deconfliction procedures to quickly adjust their use of EW or communications systems.

4-64. Electronic attack operations depend on EW support and signals intelligence to provide targeting information and battle damage assessment. However, EW officers must keep in mind that not all intelligence collection is focused on supporting EW. If not properly coordinated with the G-2 or S-2 staff, electronic attack operations may impact intelligence collection by jamming or inadvertently interfering with a particular frequency being used to collect data on the threat, or by jamming a given enemy frequency or system that deprives friendly forces of that means of collecting data. Either can significantly deter intelligence collection efforts and their ability to answer critical information requirements. Coordination between the EW officer, the fire support coordinator, and the G-2 or S-2 is prevents this interference. In situations where a known conflict between the intelligence collection effort and the use of electronic attack exists, the EW working group brings the problem to the G-3 or S-3 for resolution.

4-65. Other forms of effects rely on electromagnetic spectrum. For example, psychological operations may plan to use a given set of frequencies to broadcast messages, or a military deception plan may include the broadcast of friendly force communications. In both examples, the use of electronic attack could unintentionally interfere or disrupt such broadcasts if not properly coordinated. To ensure electronic attack does not negatively impact planned operations, the EW officer coordinates between fires, network operations, and other functional or integrating cells as required.

4-66. Like any other form of electromagnetic radiation, electronic attack can adversely affect local media and communications systems and infrastructure. EW planners consider unintended consequences of EW operations and deconflict these operations with the various functional or integrating cells. For example, friendly jamming could potentially deny the functioning of essential services such as ambulance or fire fighters to a local population. EW officers routinely synchronize electronic attack with the other functional or integrating cells responsible for the information tasks. In this way, they ensure that electronic attack efforts do not cause fratricide or unacceptable collateral damage to their intended effects.

4-67. The potential for hostile intelligence collection also affects electronic attack. A well-equipped enemy can detect friendly EW capabilities and thus gain intelligence on friendly force intentions. For example, the frequencies Army forces jam could indicate where they believe the enemy's capabilities lie. The EW officer and the G-2 or S-2 develop an understanding of the enemy's collection capability. Along with the red team (if available), they determine what the enemy might gain from friendly force use of electronic attack. (A *red team* is an organizational element comprised of trained and educated members that provide an independent capability to fully explore alternatives in plans and operations in the context of the operational environment and from the perspective of adversaries and others. [JP 2-0])

4-68. The effects of jamming only persist as long as the jammer itself is emitting and is in range to affect the target. Normally this time frame is a matter of seconds or minutes, which makes the timing of such missions critical. This is particularly true when jamming is used in direct support of aviation platforms. For example, in a mission that supports suppression of enemy air defense, the time on target and duration of the jamming must account for the speed of attack of the aviation platform. They must also account for the potential reaction time of enemy air defensive countermeasures. The development of directed-energy weapons may change this dynamic in the future. However, at present (aside from antiradiation missiles), the effects of jamming are less persistent than effects achieved by other means.

Chapter 4

ELECTRONIC PROTECTION CONSIDERATIONS

4-69. Electronic protection is achieved through physical security, communications security measures, system technical capabilities (such as frequency hopping and shielding of electronics), spectrum management, and emission control procedures. The EW officer and EW working group members must consider the following key functions when planning for electronic protection operations:

- Vulnerability analysis and assessment.
- Monitoring and feedback.
- Electronic protection measures and how they affect friendly capabilities.

Vulnerability Analysis and Assessment

4-70. Vulnerability analysis and assessment forms the basis for formulating electronic protection plans. The Defense Information Systems Agency operates the Vulnerability Analysis and Assessment Program, which specifically focuses on automated information systems and can be very useful in this effort.

Monitoring and Feedback

4-71. The National Security Agency monitors communications security. Their programs focus on telecommunications systems using wire and electronic communications. Their programs can support and remediate the command's communications security procedures when required.

Electronic Protection Measures and Their Effect on Friendly Capabilities

4-72. Electronic protection measures include any measure taken to protect the force from hostile electronic attack actions. However, these measures can also limit friendly capabilities or operations. For example, denying frequency usage to counter-radio-controlled improvised-explosive-device EW systems on a given frequency to preserve it for a critical friendly information system could leave friendly forces vulnerable to certain radio-controlled improvised explosive devices. The EW officer and the G-6 or S-6 carefully consider these second-order effects when advising the G-3 or S-3 regarding electronic protection measures.

ELECTRONIC WARFARE SUPPORT CONSIDERATIONS

4-73. The distinction between whether a given asset is performing a signals intelligence or EW support mission is determined by who tasks and controls the assets, what they are tasked to provide, and the purpose for which they are tasked. Operational commanders task assets to conduct EW support for the purpose of immediate threat recognition, targeting, planning the conduct of future operations, and other tactical actions (such as threat avoidance and homing). The EW officer coordinates with the G-2 or S-2 to ensure all EW support needed for planned EW operations is identified and submitted to the G-3 or S-3 for approval by the commander. This ensures that the required collection assets are properly tasked to provide the EW support. In cases where planned electronic attack actions may conflict with the G-2 or S-2 intelligence collection efforts, the G-3, S-3, or commander decides which has priority. The EW officer and the G-2 or S-2 develop a structured process within each echelon for conducting this intelligence gain-loss calculus during mission rehearsal exercises and predeployment work-ups.

ELECTRONIC WARFARE REPROGRAMMING CONSIDERATIONS

4-74. Electronic warfare reprogramming refers to modifying friendly EW or target sensing systems in response to validated changes in enemy equipment and tactics or the electromagnetic environment. (See paragraph 1-40 for the complete definition.) Reprogramming EW and target sensing system equipment falls under the responsibility of each Service or organization through its respective EW reprogramming support programs. It includes changes to self-defense systems, offensive weapons systems, and intelligence collection systems. During joint operations, swift identification and reprogramming efforts are critical in a rapidly evolving hostile situation. The key consideration for EW reprogramming is joint coordination. Joint coordination of Service reprogramming efforts ensures reprogramming requirements are identified, processed, and implemented consistently by all friendly forces. During joint operations, EW reprogramming coordination and monitoring is the responsibility of the joint force commander's EW staff. (For more information on EW reprogramming, see FM 3-13.10).

SECTION II — ELECTRONIC WARFARE PREPARATION

4-75. *Preparation* consists of activities performed by units to improve their ability to execute an operation. Preparation includes, but is not limited to, plan refinement; rehearsals; intelligence, surveillance, and reconnaissance; coordination; inspections; and movement (FM 3-0). Preparation creates conditions that improve friendly forces' opportunities for success. It facilitates and sustains transitions, including those to branches and sequels.

4-76. During preparation, the EW officer and members of the EW working group focus their actions on the following activities:

- Revising and refining the EW estimate, EW tasks supporting command and control warfare, and EW support to the overall plan.
- Rehearsing the synchronization of EW support to the plan (including integration into the targeting process, request procedures for joint assets, deconfliction procedures, and asset determination and refinement).
- Synchronizing the collection plan and intelligence synchronization matrix with the attack guidance matrix and EW input to the operation plan or order annexes and appendixes.
- Assessing the planned task organization developed to support EW operations, including liaison officers and organic and nonorganic capabilities required by echelon.
- Coordinating procedures with ISR operational elements (such as signals intelligence staff elements).
- Training the supporting staff members of the EW working group during mission rehearsal exercises.
- Completing precombat checks and inspections of EW assets.
- Completing sustainment preparations for EW assets.
- Coordinate with the G-4 or S-4 to develop EW equipment reporting formats.
- Completing briefbacks by subordinate EW working groups on planned EW operations.
- Refining content and format for the EW officer's portion of the battle update assessment and brief.

SECTION III — ELECTRONIC WARFARE EXECUTION

4-77. *Execution* is putting the plan into action by applying combat power to accomplish the mission and using situational understanding to assess progress and make execution and adjustment decisions (FM 3-0). Commanders focus their subordinates on executing the concept of operations by issuing their intent and mission orders.

4-78. During execution, the EW officer and EW working group members—

- Serve as the EW expert for the commander.
- Maintain the running estimate for EW operations.
- Monitor EW operations and recommend adjustments during execution.
- Recommend adjustments to the commander's critical information requirements based on the situation.
- Recommend adjustments to EW-related control measures and procedures.
- Maintain direct liaison with the fires and network operations cells and the command and control warfare working group (if formed) to ensure integration and deconfliction of EW operations.
- Coordinate and manage EW taskings to subordinate units or assets.
- Coordinate requests for nonorganic EW support.

Chapter 4

- Continue to assist the targeting working group in target development and recommend targets for attack by electronic attack assets.
- Receive, process, and coordinate subordinate requests for EW support during operations.
- Receive and process immediate support requests for suppression of enemy air defense or EW from joint or multinational forces; coordinate through fire support officer and fire support coordinator with the battlefield coordination detachment and joint or multinational liaisons for support request.
- Coordinate with airspace control section on all suppression of enemy air defense or EW missions.
- Provide input to the overall assessment regarding effectiveness of electronic attack missions.
- Maintain, update, and distribute the status of EW assets.
- Validate and disseminate cease-jamming requests.
- Coordinate and expedite electromagnetic interference reports with the analysis and control element for targeting and the spectrum manager for potential deconfliction.
- Perform jamming control authority function for ground-based EW within the assigned area of operations (when designated by the jamming control authority).

SECTION IV — ELECTRONIC WARFARE ASSESSMENT

4-79. *Assessment* is the continuous monitoring and evaluation of the current situation, particularly the enemy, and progress of an operation (FM 3-0). Commanders, assisted by their staffs, continuously assess the current situation and progress of the operation and compare it with the concept of operations, mission, and commander's intent. Based on their assessment, commanders direct adjustments, ensuring that the operation remains focused on the mission and commander's intent.

4-80. As depicted in figure 4-5 (page 4-10), assessment occurs throughout every operations process activity and includes three major tasks:
- Continuously assessing the enemy's reactions and vulnerabilities.
- Continuously monitoring the situation and progress of the operation towards the commander's desired end state.
- Evaluating the operation against measures of effectiveness and measures of performance.

4-81. The EW officer and supporting members of the EW working group make assessments throughout the operations process. During planning and preparation activities, assessments of EW are made during the MDMP, IPB, targeting, ISR synchronization, and composite risk management integration.

4-82. The EW officer, in conjunction with the G-5 or S-5, helps develop the measures of performance and measures of effectiveness for evaluating EW operations during execution. A *measure of performance* is a criterion used to assess friendly actions that is tied to measuring task accomplishment (JP 3-0). A *measure of effectiveness* is a criterion used to assess changes in system behavior, capability, or operational environment that is tied to measuring the attainment of an end state, achievement of an objective, or creation of an effect (JP 3-0). In the context of EW, an example of a measure of performance is the percentage of known enemy command and control nodes targeted and attacked by electronic attack means (action) versus the number of enemy command and control nodes that were actually destroyed or rendered inoperable for the desired duration (task accomplishment). Measures of effectiveness are used to determine the degree to which an EW action achieved the desired result. This is normally measured through analysis of data collected by both active and passive means. For example, effectiveness is measured by using radar or visual systems to detect changes in enemy weapons flight and trajectory profiles.

4-83. During execution, the EW officer and members of the EW working group participate in combat assessments within the fires cell to determine the effectiveness of electronic attack employment in support of operations. Combat assessment consists of three elements: munitions effects assessment, battle damage assessment, and reattack recommendations. (Paragraphs 4-47 to 4-49 discuss combat assessment.)

SUMMARY

4-84. The EW officer and staff members supporting the EW working group ensure the successful integration of EW capabilities into operations. The EW officer leads the EW integration effort throughout the operations process. The EW officer must be familiar with and participate in the applicable integrating processes and continuing activities discussed within this chapter.

Chapter 5
Coordination, Deconfliction, and Synchronization

Once the commander approves an operation plan or order and preparations are complete, the electronic warfare officer and supporting staff turn to coordinating, deconflicting, and synchronizing the electronic warfare efforts. They ensure electronic warfare actions are carried out as planned or are modified in response to current operations. This chapter discusses major areas and activities that require continuous coordination, deconfliction, and synchronization by the electronic warfare officer and supporting staff of the electronic warfare working groups.

COORDINATION AND DECONFLICTION

5-1. A certain amount of coordination is part of the planning process. However, once a plan is approved and an operation begins, the electronic warfare (EW) staff effort shifts to the coordination and deconfliction necessary to ensure units carry out EW actions as planned or modify actions to respond to the dynamics of the operation.

5-2. The EW officer and members of the EW working group continuously monitor several key areas. These include EW coordination across organizations (higher, lower, and adjacent units), support request coordination, electromagnetic spectrum management, EW asset management, functional coordination between EW subdivisions, EW reprogramming, and EW deconfliction. Normally, EW personnel on watch in the operations center monitor and coordinate activities of these key areas. They alert the EW officer or other EW support personnel to address the required actions.

COORDINATION ACROSS ORGANIZATIONS

5-3. At the joint level, the information operations division of the J-3 performs EW coordination. The EW section of the information operations staff engages in all EW functions. This section performs peacetime contingency planning, completes day-to-day planning and monitoring of routine theater EW activities, and crisis action planning for contingencies as part of emergent joint operations. The EW section coordinates closely with other appropriate staff sections and other larger joint planning groups as required. (JP 3-13.1 discusses joint EW coordination.)

5-4. In the early stages of contingencies, the joint force commander's EW staff assesses the staffing requirements for planning and execution. This staff also coordinates EW planning and course of action development with the joint force commander's components. Services begin component EW planning and activate their EW working groups per combatant command or Service guidelines. When the scope of a contingency becomes clearer, the command EW officer may request that the joint force commander establish a joint EW coordination cell. If a joint EW coordination cell is formed, it normally requires additional augmentation from the Service or functional components. Depending on the size of the force, EW personnel from the division, corps, or theater are expected to augment the joint EW coordination cell to form a representative EW planning and execution organization. The senior Army organization's staff EW officer anticipates this requirement and prepares to support the augmentation if requested.

5-5. Coordination occurs through established EW working groups from theater level to battalion level. Within Army organizations, the coordination of EW activities occurs both horizontally and vertically. At every level, the staff EW officer ensures the necessary coordination. Normally, coordination of EW activities between the Army and joint force air component commander flows through the battlefield coordination detachment at the joint air operations center. EW staffs at higher echelons monitor EW-related activities and resolve conflicts when necessary.

5-6. Normally the senior Army headquarters (ARFOR) G-3 or S-3 coordinates with external EW organizations, unless direct liaison is authorized at lower echelons. Other components requesting Army EW support coordinate their support requirements with the EW officer located at the ARFOR headquarters or tactical operations center. Often, a liaison from the requesting organization completes these requests. If other Service or functional components have an immediate need for Army EW support, they send the request to the operational fires directorate or fires cell and the senior headquarters EW working group (sometimes referred to as an EW coordination cell) via the Global Command and Control System or Global Command and Control System-Army. In support of external EW coordination, the staff EW officer within the J-3, G-3, or S-3—

- Provides an assessment of EW capabilities to other component operation centers.
- Coordinates preplanned EW operations with other Service components (within prescribed time lines).
- Updates preplanned EW operations in coordination with other components as required.

SUPPORT REQUEST COORDINATION

5-7. Units requesting electronic attack support forward requests to the appropriate EW working group. (See appendix D for the electronic attack request format.) Each EW working group prioritizes the requests and forwards them to the higher headquarters. The commander who owns the capability when the requested support is needed approves the requests. The technical data required to support the execution of the request is passed through EW channels at the appropriate level of classification.

5-8. Electronic warfare support requests are prioritized and passed from the EW working groups through G-2 or S-2 channels and are approved by the commander who owns the capability. New EW support requests are integrated into the intelligence synchronization process. If they are approved, they appear in the intelligence synchronization plan and the unit intelligence, surveillance, and reconnaissance plan. See FMI 2-01 for details on the intelligence synchronization process. The technical data required to support EW support requests passes via signals intelligence channels within the G-2 or S-2 by classified means.

ELECTROMAGNETIC SPECTRUM MANAGEMENT

5-9. The electromagnetic spectrum is a finite resource. Once apportioned, this resource must be managed efficiently to maximize the limited spectrum allocated to support military operations. Electromagnetic spectrum operations aim to enable electronic systems to perform their functions in the intended environment without causing or experiencing unacceptable interference. Electromagnetic spectrum operations deconflict all military, national, and host-nation systems being used in the area of operations, including electronic protection systems, communications systems, sensors, and weapon systems.

5-10. Spectrum management involves planning, coordinating, and managing use of the electromagnetic spectrum through operational, engineering, and administrative procedures. Primarily, it involves determining what specific activities will occur in each part of the available spectrum. For example, some frequencies are assigned to the counter radio-controlled improvised-explosive-device EW systems operating in the area of operations. These frequencies then are deconflicted with ground tactical communications. The spectrum manager ensures all necessary functions that require use of the electromagnetic spectrum have sufficient allocation of that spectrum to accomplish their purpose. Where a conflict (two or more functions require the same portion of the spectrum) exists, the spectrum manager resolves the conflict through direct coordination. Figure 5-1 shows the basic procedures the spectrum manager follows to deconflict spectrum use.

Coordination, Deconfliction, and Synchronization

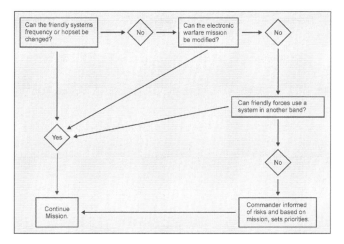

Figure 5-1. Spectrum deconfliction procedures

5-11. The spectrum manager is a member of the G-6 or S-6 section that has staff responsibility for spectrum management in the unit. The spectrum manager is a member of the unit's EW working group. Conflicts regarding spectrum use and allocation that cannot be resolved through direct coordination by the spectrum manager are referred to the G-3 or S-3 for resolution.

JAMMING CONTROL AUTHORITY

5-12. Depending on the operational situation, an Army headquarters may be designated as the jamming control authority. This authority serves as the senior jamming control authority in the area of operations. It establishes guidance for jamming on behalf of the joint force commander. If designated as the jamming control authority, the senior staff EW officer normally is tasked with the following responsibilities:
- Participating in development of and ensuring compliance with the joint restricted frequency list.
- Validating and approving or denying cease-jamming requests.
- Maintaining situational awareness of all jamming-capable systems in the area of operations.
- Acting as the joint force commander's executive agent for developing EW intelligence gain-or-loss recommendations when electronic attack or electronic warfare support conflicts occur.
- Coordinating jamming requirements with joint force components.
- Investigating unauthorized jamming events and implementing corrective measures.

See JP 3-13.1 for further information on jamming control authority.

ASSET MANAGEMENT

5-13. Regardless of echelon, the EW officer monitors and tracks the organization's EW assets and their status. The EW officer makes recommendations to the G-3 or S-3 concerning EW asset allocation and reallocation when required. The EW officer monitors and tracks EW asset status within the EW working group and reports this information to higher echelons via the Army battle command system.

OTHER COORDINATING ACTIONS

5-14. In addition to the functional considerations listed in chapter 4, several coordinating actions must also take place between the EW working groups (at all echelons) and the other planning and execution cells within the headquarters. These actions include—

- Detailed coordination between the EW activities and the intelligence activities supporting an operation.
- Coordination of EW systems reprogramming.
- Coordination with the working groups or cells coordinating the command and control warfare and information protection tasks.

Coordination Between EW Activites and Intelligence Activities

5-15. Most of the intelligence effort, before and during an operation, relies on collection activities targeted against various parts of the electromagnetic spectrum. Electronic warfare support depends on the timely collection, processing, and reporting of intelligence and combat information to alert EW operators and other military activities about intelligence collected in the electromagnetic spectrum. The EW officer and G-2 or S-2 ensure EW collection priorities and EW support collection assets are integrated into a complete intelligence collection plan. This plan ensures that units maximize the use of scarce intelligence and collection assets to support the commander's objectives.

Coordination of EW Systems Reprogramming

5-16. The EW officer and G-2, at division and corps levels, track and coordinate EW systems reprogramming input submitted by lower echelons. This input is then forwarded to the Army Service component command headquarters for submission to the Army Reprogramming Analysis Team. EW officers ensure this input is promptly submitted to ensure urgent reprogramming actions are completed for assigned systems. See FM 3-13.10 for detailed procedures for reprogramming EW and target sensing systems.

Coordination Between EW, Command and Control Warfare, and Information Tasks

5-17. EW working groups coordinate their supporting actions with the elements responsible for the Army information tasks—information engagement, command and control warfare, information protection, operations security, and military deception. Although EW plays a major role in supporting command and control warfare and information protection, it also enhances or provides direct support to other information tasks. For example, enemy radio and television broadcasts can be disrupted or replaced with friendly radio and television messages as part of larger psychological operations in support of information engagement. Electronic deception capabilities can support and enhance an overall military deception operation.

DECONFLICTION

5-18. Friendly forces depend on electromagnetic energy and the electromagnetic spectrum to sense, process, store, measure, analyze, and communicate information. This dependency creates the potential for significant interference between various friendly systems. Without proper deconfliction, interference could damage friendly capabilities or lead to operational failure. This is especially true with regard to EW systems. EW deconfliction includes—

- Friendly electromagnetic spectrum use for communications and other purposes (such as navigation systems and sensors) with electronic attack activities (such as counter-radio-controlled improvised-explosive-device EW systems).
- Electronic attack activities with electronic warfare support activities (potential electromagnetic interference of collection assets).

- Electronic attack and electronic warfare support activities with information tasks involving electromagnetic emissions (such as counter-radio-controlled improvised-explosive-device EW systems interfering with a psychological-operations radio broadcast).
- Electronic attack activities with host-nation electromagnetic spectrum users (such as commercial broadcasters, emergency first responders, and law enforcement).

5-19. The forum for deconfliction is the unit's EW working group. As such, the specific composition of the working group may expand to include more than the standard staff representation described in chapter 3. Regardless of echelon, to perform its critical deconfliction function, the EW working group retains knowledgeable representation from and ready access to decisionmakers. The EW working group also retains knowledge of and access to higher headquarters assistance and reachback capabilities available (See appendix F for more information).

SYNCHRONIZATION

5-20. EW, particularly in electronic attack, can produce both intended and unintended effects. Therefore, units thoroughly synchronize its use with other forms of fires and with friendly systems operating in the electromagnetic spectrum. Through synchronization, units avoid negative effects such as communications fratricide by jammers. The EW officer ensures all EW activities are integrated into the appropriate sections of plans—fires, information protection, command and control warfare, and military deception plans. This officer also synchronizes EW activities for maximum contribution to the commander's desired effects while preventing EW from inhibiting friendly force capabilities. The primary forum for this synchronization is the unit's EW working group. The EW officer attends the regular targeting meetings in the fires cell and may also participate (perhaps as a standing member) in other functional or integrating cells and working groups. These may include fires, information engagement, network operations, or future operations. The EW officer's participation in these other cells and working groups helps to synchronize EW operations.

SUMMARY

5-21. EW capabilities yield many advantages for the commander. The EW working group's sole purpose is to facilitate the integration, coordination, deconfliction, and synchronization of EW operations to ensure advantages are achieved. This effort requires constant coordination with the unit's other functional cells and working groups. As conflicts are identified during the planning and execution of operations, the EW officer and supporting staff members coordinate solutions to those conflicts within the EW working group.

Chapter 6
Integration with Joint and Multinational Operations

Joint warfare is team warfare. It requires the integrated and synchronized application of all appropriate capabilities. During joint operations, Services work together to accomplish a mission. In multinational operations, forces of two or more nations work together to accomplish a mission. During both joint and multinational operations, forces operate under established organizational frameworks and coordination guidelines. This chapter describes the joint and multinational operational frameworks and guidelines for integrating electronic warfare capabilities.

JOINT ELECTRONIC WARFARE OPERATIONS

6-1. One strength of operating as a joint force is the ability to maximize combat capabilities through unified action. However, the ability to maximize the capabilities of a joint force requires guidelines and an organizational framework that can be used to integrate them effectively. JP 3-13.1 establishes the guidelines and organizational framework for joint electronic warfare (EW) operations.

6-2. Joint task forces are task-organized. Therefore, their composition varies based on the mission. Normally the EW organization within a joint force centers on the—
- Component commands.
- Supporting joint centers.
- Joint force staff.
- Joint force commander's EW staff, joint electronic warfare coordination cell, or information operations (IO) cell.

The supporting centers for EW operations may include the joint operations center, joint intelligence center, Joint Frequency Management Office (JFMO), and joint targeting coordination board.

JOINT FORCE PRINCIPAL STAFF FOR ELECTRONIC WARFARE

6-3. In EW, the principal staff consists of the J-2, J-3, and J-6. The J-2 collects, processes, tailors, and disseminates all-source intelligence for EW. The J-3 has primary staff responsibility for EW activity. This director also plans, coordinates, and integrates joint EW operations with other combat disciplines in the joint task force. Normally, the joint force commander's EW staff or a joint EW coordination cell and an IO cell assist the J-3. The joint force staff network operations director (in the J-6) coordinates electromagnetic spectrum use for information systems with electromagnetic-dependent weapons systems used by the joint force. The IO officer is the principal IO advisor to the J-3. This officer is the lead planner for integrating, coordinating, and executing IO. The command EW officer is the principal EW planner on the J-3 staff. This officer coordinates with the IO cell to integrate EW operations fully with other IO core, supporting, and related capabilities (see JP 3-13.1 for further information)

JOINT FORCE COMMANDER'S ELECTRONIC WARFARE STAFF

6-4. A joint force commander's EW staff supports the joint force commander in planning, coordinating, synchronizing, and integrating joint force EW operations. The joint force commander's EW staff ensures that joint EW capabilities support the joint force commander's objectives. The joint force commander's EW staff is an element within the J-3. It consists of representatives from each component of the joint force.

An EW officer appointed by the J-3 leads this element. The joint force commander's EW staff includes representatives from the J-2 and J-6 to facilitate intelligence support and EW frequency deconfliction.

6-5. On many joint staffs, the intra-staff coordination previously accomplished through a joint force commander's EW staff is now performed by an IO cell or similar organization. An IO cell, if established, coordinates EW activities with other IO activities to maximize effectiveness and prevent mutual interference. If both a joint force commander's EW staff and an IO cell exist, a joint force commander's EW staff representative may be assigned to the IO cell to facilitate coordination. For more information about the organization and procedures of the joint IO cell, see JP 3-13.

JOINT ELECTRONIC WARFARE COORDINATION CELL

6-6. The decision to form a joint EW coordination cell depends on the anticipated role of EW in an operation. When EW is expected to play a significant role in the joint force commander's mission, a component command's EW coordination organization may be designated as the joint EW coordination cell to handle the EW aspects of the operation. The joint EW coordination cell may be part of the joint force commander's staff, be assigned to the J-3 directorate, or remain within the designated component commander's structure. The joint EW coordination cell plans operational-level EW for the joint force commander. (JP 3-13.1 discusses the joint EW coordination cell in more detail.)

JOINT TASK FORCE COMPONENT COMMANDS

6-7. Joint task force component commanders exercise operational control of their EW assets. Each component is organized and equipped to perform EW tasks in support of its basic mission and to provide support to the joint force commander's overall objectives. If a component command (Service or functional) is designated to stand up a joint EW coordination cell, it executes the responsibilities and functions outlined in JP 3-13.1.

6-8. A major consideration for standing up a joint EW coordination cell at the component command level is access to a special compartmented information facility to accomplish the cell's required coordination functions. Optimal joint EW coordination cell staffing dictates including special technical operations personnel cleared to coordinate and deconflict special technical operations issues. Special technical operations are associated with the planning and coordination of advanced special programs and the integration of new capabilities into operational units.

6-9. Under current force structure, the special technical operations requirement limits the activation of a joint EW coordination cell to organizations at corps and above levels. Organizations below corps level require significant joint augmentation to meet the special technical operations requirement.

JOINT FREQUENCY MANAGEMENT OFFICE

6-10. Joint policy tasks each geographic combatant commander to establish a structure to manage spectrum use and establish procedures that support ongoing operations. This structure must include a JFMO. The JFMO may be assigned from the supported combatant commander's J-6 staff, from a component's staff, or from an external command such as the Joint Spectrum Center. The JFMO coordinates the information systems use of the electromagnetic spectrum, frequency management, and frequency deconfliction. The JFMO develops the frequency management plan and makes recommendations to alleviate mutual interference.

6-11. The G-6 or S-6 coordinates the Army's use of the electromagnetic spectrum, frequency management, and frequency deconfliction with the JFMO through the network operations cell. If established, coordination with the joint spectrum management element is required. (See figure 6-1.)

Integration with Joint and Multinational Operations

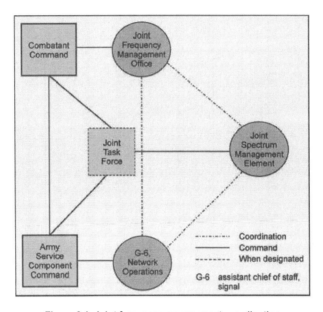

Figure 6-1. Joint frequency management coordination

JOINT INTELLIGENCE CENTER

6-12. The joint intelligence center is the focal point for the intelligence structure supporting the J-2. Directed by the J-2, the joint intelligence center communicates directly with component intelligence agencies and monitors intelligence support to EW operations. This center can adjust intelligence gathering to support EW missions. Within the G-2, EW support requests are coordinated through the requirement cell and then forwarded to the requirements division within the joint intelligence center. (See figure 6-2, page 6-4.)

Chapter 6

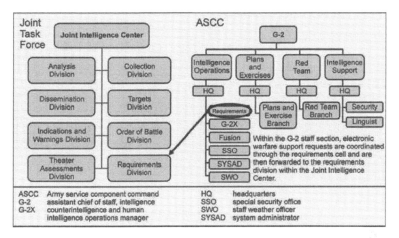

Figure 6-2. Electronic warfare support request coordination

6-13. The composition and focus of each joint intelligence center varies by theater. However, each can perform indications and warning as well as collect, manage, and disseminate current intelligence. Through the joint intelligence center, the ARFOR (Army Service component) headquarters coordinates support from the Air Force, Navy, and Marine Corps and national, interagency, and multinational sources. In addition to its other functions, the joint intelligence center coordinates the acquisition of national intelligence for the joint task force and the combatant command's staff.

JOINT TARGETING COORDINATION BOARD

6-14. The joint targeting coordination board focuses on developing broad targeting priorities and other targeting guidance in accordance with the joint force commander's objectives as they relate operationally. The joint targeting coordination board remains flexible enough to address targeting issues without becoming overly involved in tactical-level decisionmaking. Briefings conducted at the joint targeting coordination board focus on ensuring that intelligence, operations (by all components and applicable staff elements), fires, and maneuver are on track, coordinated, and synchronized. For further information on the joint targeting coordination board, see JP 3-60.

MULTINATIONAL ELECTRONIC WARFARE OPERATIONS

6-15. EW is an integral part of multinational operations (sometimes referred to as combined operations). U.S. planners integrate U.S. and multinational EW capabilities into a single, integrated EW plan. U.S. planners provide multinational forces with information concerning U.S. EW capabilities and provide them EW planning and operational support. However, the planning of multinational force EW is difficult due to security issues, differences in levels of training, language barriers, and terminology and procedural issues. U.S. and North Atlantic Treaty Organization (NATO) EW doctrine provide commonality and a framework for using EW in NATO operations. (See Allied Joint Publication 3.6 for specific information.)

MULTINATIONAL FORCE COMMANDER

6-16. The multinational force commander provides guidance for planning and conducting EW operations to the multinational force through the C-3 and the EW coordination cell. The EW coordination cell is located at multinational force headquarters. An IO cell may also be established to coordinate all IO-related activities, including related EW operations.

JOINT OPERATIONS STAFF SECTION

6-17. Within the multinational staff, the joint operations section has primary responsibility for planning and integrating EW activities. A staff EW officer is designated with specific responsibilities. These include integrating multinational augmentees, interpreting or translating EW plans and procedures, coordinating appropriate communications connectivity, and integrating multinational force communications into a joint restricted frequency list.

MULTINATIONAL ELECTRONIC WARFARE COORDINATION CELL

6-18. In multinational operations, the multinational force commander uses an EW coordination cell as the mechanism for coordinating EW resources within the area of operations. This cell is an integral part of the multinational joint force headquarters J-3 staff, at whatever level is appropriate. It provides an effective means of coordinating all EW activities by the multinational force. The multinational force EW coordination cell plans and coordinates all in-theater EW activities in close liaison with the J-2, J-5, and J-6.

ELECTRONIC WARFARE MUTUAL SUPPORT

6-19. Electronic warfare mutual support is the timely exchange of EW information to make the best use of the available resources. It is facilitated by the use of an agreed reference database called the NATO emitter database. Electronic warfare mutual support procedures developed during EW planning include—
- A review of friendly and enemy information data elements that may be exchanged.
- Mechanisms leading to the exchange of data during peace, crisis, and war.
- Development of peacetime exercises to practice the exchange of data.
- Establishment of EW points of contact with adjacent formations and higher and subordinate headquarters for planning purposes, regardless of whether EW resources exist or not.
- Initial acquisition and maintenance of multinational force EW capabilities.
- Exchange of EW liaison teams equipped with appropriate communications.
- Establishment and rehearsal of contingency plans for the exchange of information on friendly and enemy forces.
- Development of communications protocols in accordance with NATO Standardization Agreement (STANAG) 5048.
- Provision of secure, dedicated, and survivable communications.

OTHER CONSIDERATIONS

6-20. EW in multinational operations addresses other considerations. Soldiers must consider—
- Exchange of EW information.
- Exchange of signals intelligence information.
- Exchange of the electronic order of battle.
- Electronic warfare reprogramming.

6-21. Army forces participating in multinational EW operations must exchange EW information with other forces. They must help develop joint information exchange protocols and use those protocols for conducting operations.

6-22. Exchanging signals intelligence information requires care to avoid violating signals intelligence security rules. The policy and relationship between EW and signals intelligence within NATO are set out in NATO Military Committee (MC) 64.

Chapter 6

6-23. In peacetime, before forming a multinational force, the exchange of electronic order of battle information is normally achieved under bilateral agreement. During multinational operations, a representative of the joint EW coordination cell, through the theater joint analysis center or the joint intelligence center, ensures the maintenance of an up-to-date electronic order of battle. The inclusion of multinational forces is based on security and information exchange guidelines agreed upon by the participating nations.

6-24. Electronic warfare reprogramming is a national responsibility. However, the joint EW coordination cell remains aware of reprogramming efforts being conducted within the multinational force. FM 3-13.10 guides the Army's reprogramming effort.

SUMMARY

6-25. Every joint or multinational operation is uniquely organized to accomplish the mission. Army EW officers integrate EW forces and capabilities with the organizations and agencies outlined in this chapter. To coordinate Army EW operations with joint and multinational forces, Army EW officers must understand fully the organizational frameworks, policies, and guidelines established for joint and multinational EW operations.

Chapter 7
Electronic Warfare Capabilities

Electronic warfare capabilities consist of high-demand, low-density assets across the Services. Hence, the conduct of electronic warfare operations requires joint interdependence. This complex interdependence extends beyond the traditional Service capabilities. It includes national agencies—such as the Central Intelligence Agency, National Security Agency, and Defense Intelligence Agency—that constantly seek to identify, catalog, and update the electronic order of battle of enemies and adversaries. To support the joint force commander, the subject matter expertise and unique capabilities provided by each Service, agency, and branch or proponent are integrated with all available electronic warfare capabilities.

SERVICE ELECTRONIC WARFARE CAPABILITIES

7-1. Each Service maintains electronic warfare (EW) capabilities to support operational requirements. During operations, the Army is dependent on organic and nonorganic EW capabilities from higher echelons, joint forces, and national agencies. Army EW planners leverage all available EW capabilities to support Army operations. Although not all-inclusive, appendix E provides a listing of current Army, Marine Corps, Navy, and Air Force EW capabilities and references.

EXTERNAL SUPPORT AGENCIES AND ACTIVITIES

7-2. Army EW planners routinely use and receive support from external organizations to assist in planning and integrating EW operations. Support from these organizations may include personnel augmentation, functional area expertise, technical support, and planning support.

BIG CROW PROGRAM OFFICE

7-3. The Big Crow Program Office was established in 1971 to provide testing environments for U.S. military radio frequency sensor, communication, and navigation systems. Today, the Big Crow Program Office provides customers with joint, multifunctional support for testing communications, sensors, information operations, and related weapon systems in support of Department of Defense (DOD), the individual Services, the National Aeronautics and Space Administration, the National Reconnaissance Office, and others. This support includes replicating information operations and EW threat environments as well as providing telemetry recording, technology prototyping, proof-of-concept demonstrations, and information operations and EW training. Big Crow's mission and capabilities now span the electromagnetic spectrum, encompassing EW, telemetry, radar, and electro-optical systems. Mobile and worldwide deployable, the Big Crow Program Office offers a variety of capabilities.

DEFENSE INFORMATION SYSTEMS AGENCY

7-4. The Defense Information Systems Agency is a combat support agency. It plans, develops, fields, operates, and supports command, control, communications, and information systems. These systems serve the President, the Secretary of Defense, the Joint Chiefs of Staff, the combatant commanders, and other DOD components. The Defense Information Systems Agency also operates the Vulnerability Analysis and Assessment Program. This program specifically focuses on automated information systems.

Chapter 7

JOINT COMMUNICATIONS SECURITY MONITOR ACTIVITY

7-5. The Joint Communications Security Monitor Activity was created in 1993 by a memorandum of agreement between the Services' operations deputies, Directors of the Joint Staff, and the National Security Agency. The Joint Communications Security Monitor Activity monitors (collects, analyzes, and reports) communications security of DOD telecommunications and automated information systems as well as related noncommunications signals. Its purpose is to identify potentially exploitable vulnerabilities and to recommend countermeasures and corrective actions. The Joint Communications Security Monitor Activity supports real world operations, joint exercises, and DOD systems monitoring.

JOINT INFORMATION OPERATIONS WARFARE COMMAND

7-6. The Joint Information Operations Warfare Command (JIOWC) was activated in 2006 as a functional component to the United States Strategic Command (USSTRATCOM). JIOWC integrates joint information operations into military plans, exercises, and operations across the spectrum of conflict. It is a valuable resource for commanders during the planning and execution of joint information operations. JIOWC deploys information operations planning teams when the commander of USSTRATCOM approves a request for support. This center delivers tailored, highly skilled support and sophisticated models and simulations to joint commanders and provides information operations expertise in joint exercises and contingency operations.

7-7. JIOWC also fields the Joint Electronic Warfare Center. This center provides specialized expertise in EW. It is an innovation center for existing and emerging EW capabilities and tactics, techniques, and procedures via a network of units, labs, test ranges, and academia. The Joint Electronic Warfare Center also has EW reprogramming oversight responsibilities for the Joint Staff. This oversight includes organizing, managing, and exercising joint aspects of EW reprogramming and facilitating the exchange of joint EW reprogramming data. The actual reprogramming of equipment, however, is a Service responsibility.

JOINT SPECTRUM CENTER

7-8. The Joint Spectrum Center was activated in 1994 under the direction of the joint staff's J-6. The Joint Spectrum Center assumed all the missions and responsibilities previously performed by the Electromagnetic Compatibility Center plus additional responsibilities. Personnel in the Joint Spectrum Center are experts in spectrum planning, electromagnetic compatibility and vulnerability, electromagnetic environmental effects, information systems, modeling and simulation, operations support, and system acquisition. The Joint Spectrum Center provides complete, spectrum-related services to combatant commanders, Services, and other government agencies. The Joint Spectrum Center deploys teams in support of the combatant commanders and serves as the DOD focal point for supporting spectrum supremacy aspects of information operations. It assists Soldiers in developing and managing the joint restricted frequency list and helps to resolve operational interference and jamming incidents. The Joint Spectrum Center can also provide databases of friendly force command and control systems for use in planning electronic protection. The Joint Spectrum Center is a field office within the Defense Spectrum Organization under the Defense Information Systems Agency.

JOINT WARFARE ANALYSIS CENTER

7-9. The Joint Warfare Analysis Center is a Navy-sponsored joint command under the J-3 established in 1994. The Joint Warfare Analysis Center assists the Chairman of the Joint Chiefs of Staff and combatant commanders in preparing and analyzing joint operational plans. It provides analysis of engineering and scientific data and integrates operational analysis with intelligence.

MARINE CORPS INFORMATION TECHNOLOGY AND NETWORK OPERATIONS CENTER

7-10. The Marine Corps Information Technology and Network Operations Center is the Marine Corps' enterprise network operations center. The Marine Corps Information Technology and Network Operations Center is the nerve center for the central operational direction and configuration management of the Marine

Corps enterprise network. It is co-located with the Marine Corps forces computer network defense, the component to the joint task force for computer network operations, and the Marine Corps computer incident response team. This relationship provides a strong framework for integrated network management and defense.

NATIONAL SECURITY AGENCY

7-11. The National Security Agency/Central Security Service is America's cryptologic organization. This organization protects U.S. government information systems and produces foreign signals intelligence information. Executive Order 12333, 4 December 1981, describes the responsibility of the National Security Agency/Central Security Service in more detail. The resources of National Security Agency/Central Security Service are organized for two national missions:

- The Information Assurance Mission provides the solutions, products, and services, and conducts defensive information operations, to achieve information assurance for information infrastructures critical to U.S. national security interests.
- The Signals Intelligence Mission allows for an effective, unified organization and control of all the foreign signals collection and processing activities of the United States. The National Security Agency is authorized to produce signals intelligence in accordance with objectives, requirements, and priorities established by the Director of National Intelligence in consultation with the President's Foreign Intelligence Advisory Board.

7-12. The Director, National Security Agency is the principal signals intelligence and information security advisor to the Secretary of Defense, Director of National Intelligence, and the Chairman of the Joint Chiefs of Staff. The Director, National Security Agency provides signals intelligence support to combatant commanders and others in accordance with their expressed formal requirements.

SUMMARY

7-13. This chapter and appendix E provide a sampling of available joint and Service EW capabilities, activities, and agencies that support ground force commanders in full spectrum operations. To leverage these capabilities for EW support, Army EW officers acquire a working knowledge of the capabilities available and the procedures for requesting support. Additionally, appendix F provides information on available EW related tools and other resources.

Appendix A
The Electromagnetic Environment

Electromagnetic energy is both a natural and manmade occurrence. This energy, in the form of electromagnetic radiation, consists of oscillating electric and magnetic fields and is propagated at the speed of light. Electromagnetic radiation is measured by the frequency of its wave pattern's repetition within a set unit of time. The standard term for the measurement of electromagnetic radiation is the hertz (Hz), the number of repetitions (cycles) per second. The electromagnetic spectrum refers to the range of frequencies of electromagnetic radiation.

OVERVIEW OF THE ELECTROMAGNETIC ENVIRONMENT

A-1. The electromagnetic environment is the resulting product of the power and time distribution, in various frequency ranges, of radiated or conducted electromagnetic emission levels. Within their intended operational environment, a military force, system, or platform may encounter these emissions while performing tasks during operations. The electromagnetic environment is the sum of—

- Electromagnetic interference.
- Electromagnetic pulse.
- Hazards of electromagnetic radiation to personnel, ordnance, and volatile materials.
- Natural phenomena effects of lightning and precipitation static. (*Precipitation static* is charged precipitation particles that strike antennas and gradually charge the antenna, which ultimately discharges across the insulator, causing a burst of static [JP 3-13.1]).

Appendix A

THE ELECTROMAGNETIC SPECTRUM

A-2. The *electromagnetic spectrum* is the range of frequencies of electromagnetic radiation from zero to infinity. It is divided into 26 alphabetically designated bands (JP 1-02). The spectrum is a continuum of all electromagnetic waves arranged according to frequency and wavelength. The electromagnetic spectrum extends from below the frequencies used for modern radio (at the long-wavelength end) through gamma radiation (at the short-wavelength end). It covers wavelengths from thousands of kilometers to a fraction of the size of an atom. Figure A-1 shows the spectrum regions and wavelength segments associated with the electromagnetic spectrum.

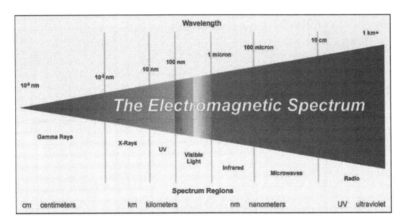

Figure A-1. The electromagnetic spectrum

A-3. Included within the radio and microwave regions of the electromagnetic spectrum are the radio frequency and radar bands. These bands are routinely referred to by their band designators. For example, high frequency radios are HF radios and K-band radars are radars that operate between 18 and 27 gigahertz. Civilian agencies and military forces throughout the world use several different designator systems, which can result in confusion. Table A-1 shows the radio frequency band designators and their associated frequency ranges. It also shows radar band designators, associated frequency ranges, and typical usage. These are standard designations used by the United States.

The Electromagnetic Environment

Table A-1. Radio and radar designators and frequency bands

Radio Frequency Band Designator	Radio Frequency Range	Radar Band Designator*	Frequency Range	Typical Usage
ULF	lower than 3 Hz	VHF	50-330 MHZ	Very long-range surveillance
ELF	3 Hz - 3 kHz	UHF	300-1,000 MHz	Very long-range surveillance
VLF	3 - 30 kHz	L	1-2 Ghz	Long-range surveillance, enroute traffic control
LF	30 - 300 kHz	S	2-4 Ghz	Moderate-range surveillance, terminal traffic control, long-range weather
MF	300 kHz - 3 MHz	C	4-8 Ghz	Long-range tracking, airborne weather
HF	3 - 30 MHz	X	8-12 Ghz	Short-range tracking, missile guidance, mapping, marine radar, airborne intercept
VHF	30 - 300 MHz	K_u	12-18 Ghz	High resolution mapping, satellite altimetry
UHF	300 MHZ - 3 GHz	K	18-27 Ghz	Little use
SHF	3 - 30 GHz	K_a	27-40 Ghz	Very high resolution mapping, airport surveillance
EHF	30 - 300 GHz			
Sub-millimeter	300 Ghz - 1 THz			

EHF	extremely high frequency	kHz	kilohertz	THz	terahertz
ELF	extremely low frequency	LF	low frequency	UHF	ultra high frequency
GHz	Gigahertz	MF	medium frequency	ULF	ultra low frequency
HF	high frequency	MHz	megahertz	VHF	very high frequency
Hz	hertz	SHF	super high frequency		

* Radar band designators relate back to the early development of radar in World War II when the letter designators were used for purposes of secrecy. After the requirement for secrecy was no longer needed, these letter band designators remained.

MILITARY OPERATIONS AND THE ELECTROMAGNETIC ENVIRONMENT

A-4. The impact of the electromagnetic environment upon the operational capability of military forces, equipment, systems, and platforms is referred to as electromagnetic environmental effects. Electromagnetic environmental effects encompass all electromagnetic disciplines, including—

- Electromagnetic compatibility and electromagnetic interference.
- Electromagnetic vulnerability.
- Electromagnetic pulse.
- Electronic protection.
- Hazards of electromagnetic radiation to personnel, ordnance, and volatile materials (such as fuels).
- Natural phenomena effects of lightning and precipitation static.

A-5. *Electromagnetic vulnerability* consists of the characteristics of a system that cause it to suffer a definite degradation (incapability to perform the designated mission) as a result of having been subjected to a certain level of electromagnetic environmental effects (JP 3-13.1). Electronic warfare support plays a key role in identifying the electromagnetic vulnerability of an adversary's electronic equipment and systems. Friendly forces take advantage of these vulnerabilities through electronic warfare operations.

Appendix A

DIRECTED ENERGY

A-6. Directed energy refers to technologies that produce of a beam of concentrated electromagnetic energy or atomic or subatomic particles (see chapter 1). *Directed-energy warfare* is military action involving the use of directed-energy weapons, devices, and countermeasures to either cause direct damage or destruction of enemy equipment, facilities, and personnel, or to determine, exploit, reduce, or prevent hostile use of the electromagnetic spectrum through damage, destruction, and disruption. It also includes actions taken to protect friendly equipment, facilities, and personnel and retain friendly use of the electromagnetic spectrum (JP 1-02). A *directed-energy weapon* is a system using directed energy primarily as a direct means to damage or destroy enemy equipment, facilities, and personnel (JP 1-02). In addition to destructive effects, directed-energy weapons can also support area denial, crowd control, and obscuration.

A-7. The application of directed energy includes lasers, radio-frequency weapons, and particle-beam weapons. As directed-energy weapons evolve, the tactics, techniques, and procedures for their use also evolve to ensure their safe, effective employment. In electronic warfare, most directed-energy applications fit into the category of electronic attack. However, other applications can be categorized as electronic protection or even electronic warfare support. Examples include the following:

- Applications used for electronic attack, which may include—
  - A laser designed to blind or disrupt optical sensors.
  - A millimeter wave directed-energy weapon used for crowd control.
  - A laser-warning receiver designed to initiate a laser countermeasure to defeat a laser weapon.
  - A millimeter wave obscuration system used to disrupt or defeat a millimeter wave system.
  - A device used to counter radio-controlled improvised explosive devices.
- A laser-warning receiver designed solely to detect and analyze a laser signal is used for electronic warfare support.
- A visor or goggle designed to filter out the harmful wavelength of laser light is used for electronic protection.

A-8. As the use of destructive directed-energy weapons grows, Army forces require the capability to collect information on them. Additionally, Army forces require tactics, techniques, and procedures to mitigate directed-energy weapon effects. Currently, the definitions and terms relating to directed energy are articulated within electronic warfare doctrine. As the technologies related to directed energy expand, joint and Army doctrine may discuss employing directed energy under other doctrinal subjects.

Appendix B
Electronic Warfare Input to Operation Plans and Orders

This appendix discusses electronic warfare input to Army and joint plans and orders.

ARMY PLANS AND ORDERS

B-1. This paragraph lists the electronic warfare (EW) information required for Army operation plans and orders. (See figure B-1 on page B-2 for the EW appendix format.) This discussion is based on current doctrine from FM 5-0. When it is republished, FM 5-0 will state where to place EW-related information in the revised plans and orders format. In addition to the appendix 4 (Electronic Warfare) to Annex P (Information Operations), the following components of operation plans and orders may require EW input:

- **Base order or plan**:
  - Sub-subparagraph (2) (Fires) to subparagraph a (Concept of Operations) to paragraph 3 (Execution).
  - Sub-subparagraph (7) (Information Operations) to subparagraph a (Concept of Operations) to paragraph 3 (Execution).
- **Annex D (Fire Support)**:
  - Sub-subparagraph (4) (Electronic Warfare) to subparagraph b (Air Support) to paragraph 3 (Execution)
  - Appendix 1 (Air Support).
- **Annex L (Intelligence, Surveillance, and Reconnaissance)**:
  - Sub-subparagraph (2) (Fires) to subparagraph a (Concept of Operations) to paragraph 3 (Execution).
  - Sub-subparagraph (7) (Information Operations) to subparagraph a (Concept of Operations) to paragraph 3 (Execution).
- **Annex N (Space)**: Sub-subparagraph (10) (Electronic Warfare) to subparagraph b (Space Activities) to paragraph 3 (Execution).
- **Annex P (Information Operations)**:
  - Sub-sub-subparagraph (d) (Electronic Warfare) to sub-subparagraph (8) to subparagraph a (Concept of Support) to paragraph 3 (Execution).
  - Sub-subparagraph (3) (List of Tasks to Electronic Warfare Units) to subparagraph b (Tasks to Subordinate Units) to paragraph 3 (Execution).

Appendix B

```
                              [Classification]

Appendix 4 (Electronic Warfare) to Annex P (Information Operations) to OPORD No____

1. SITUATION.

    a. Enemy.
        • Identify the vulnerabilities of enemy information systems and electronic warfare systems.
        • Identify the enemy capability to interfere with accomplishment of the electronic warfare mission.

    b. Friendly.
        • Identify friendly electronic warfare assets and resources that affect electronic warfare planning by
          subordinate commanders.
        • Identify friendly foreign forces with which subordinate commanders may operate.
        • Identify potential conflicts within the friendly electromagnetic spectrum, especially if conducting joint
          or multinational operations. Identify and de-conflict methods and priority of spectrum distribution.

    c. Attachments and detachments.
        • List the electronic warfare assets that are attached or detached.
        • List the electronic warfare resources available from higher headquarters.

2. MISSION. State how electronic warfare will support the commander's objectives.

3. EXECUTION.

    a. Scheme of support. State the electronic warfare tasks.

    b. Tasks to subordinate units. Identify the electronic warfare tasks for each unit.

    c. Coordinating instructions.
        • Identify electronic warfare instructions applicable to two or more units.
        • Identify the requirements for the coordination of electronic warfare actions between units.
        • Identify the emission control guidance.

4. SERVICE SUPPORT. Identify service support for electronic warfare operations.

5. COMMAND AND SIGNAL.

    a. Command.

    b. Signal. Identify if any, the special or unusual electronic warfare-related communications requirements.

                              [Classification]
```

Figure B-1. Appendix 4 (Electronic Warfare) to annex P (Information Operations) instructions

JOINT PLANS AND ORDERS

B-2. If required to provide EW input to portions of a joint order, the primary areas for input are the following:

- Paragraph 3 (Execution) to appendix 3 (Information Operations) to Annex C (Operations).
- Tab B (Electronic Warfare) to appendix 3 (Information Operations) to Annex C (Operations).

B-3. See CJCSM 3122.03C for the Joint Operations Planning and Execution System format.

Appendix C
Electronic Warfare Running Estimate

This appendix discusses the electronic warfare running estimate. A *running estimate* is a staff section's continuous assessment of current and future operations to determine if the current operation is proceeding according to the commander's intent and if future operations are supportable (FM 3-0).

C-1. The electronic warfare (EW) running estimate is used to support the military decisionmaking process during planning and execution. During planning, the EW running estimate provides an assessment of the supportability of each proposed course of action from an EW perspective. The format of the EW running estimate closely parallels the steps of the military decisionmaking process. It serves as the primary tool for recording the EW officer's assessments, analyses, and recommendations for EW operations. The EW officer and staff in the EW working group are responsible for conducting the analysis and providing recommendations based on the EW running estimate.

C-2. A complete EW running estimate should contain the information necessary to answer any question the commander may pose. If there are gaps in the EW running estimate, the staff identifies the gaps as information requirements and submits them to the intelligence cell. The EW running estimate can form the basis for EW input required in other applicable appendixes and annexes within operation plans and orders. Figure C-1 on page C-2 provides a sample EW running estimate for use during planning.

Appendix C

---

1. **MISSION.** Show the restated mission resulting from mission analysis.

2. **SITUATION AND CONSIDERATIONS.**

   a. Characteristics of the area of operations.
      - Weather. State how the weather may impact EW operations.
      - Terrain. State how aspects of the terrain may impact EW operations.
      - Civil Considerations. State how rules of engagement and civil emergency responder frequency restrictions may impact EW operations.

   b. Enemy forces. Discuss enemy dispositions, composition, strength, capabilities, and courses of action (COAs) as they affect EW operations. Identify enemy EW vulnerabilities.

   c. Friendly forces.
      - List the current status of the forces EW resources.
      - List the current status of additional EW support resources.
      - Provide a comparison of EW support requirements with available capabilities and recommend solutions for any discrepancies.
      - Identify friendly forces EW vulnerability and recommend solutions.

   d. Assumptions. List any assumptions used that may affect the employment of EW capabilities.

3. **COURSES OF ACTION.**

   a. List the friendly COAs that were wargamed.
   b. List the evaluation criteria identified during the COA analysis.

4. **ANALYSIS.** Analyze each COA using the evaluation criteria identified during COA analysis.

5. **COMPARISON.** Compare each COA. Rank order the COAs for each EW key consideration identified.

6. **RECOMMENDATION AND CONCLUSIONS.** This paragraph translates the "best" course of action (as determined in paragraph 5) into a complete recommendation. It should outline who, what, where, when, how, and why from the EW point of view. It states which course of action can best be supported by friendly EW, and is less vulnerable to enemy EW force capabilities.

   a. Recommend the most supportable COA from an EW perspective.
   b. List any EW related issues, deficiencies and risks and provide recommendations to reduce their impact.

   ANNEXES: Include annexes as required. Annexes with pertinent details should be used to the extent practical to support the contents of the estimate. These annexes may be in considerable detail with only the high points included in the body of the estimate. Annexes should add depth to the contents of the estimate, but should not be used as a substitute for key points that should be included in the body of the estimate.

Figure C-1. Example of an electronic warfare running estimate

C-3. Once the commander approves the order, the EW running estimate is used to inform current and future operations. During execution the EW running estimate is used to help determine if current EW operations are proceeding according to plan and if future EW operations are supportable. Figure C-2, page C-3, shows a sample of the information that might be used to update the EW running estimate during execution. The EW officer and supporting staff members within the EW working group produce and update the running estimate.

Current operation order and fragmentary orders
- Define the battlefield environment. Focus on the aspects of the terrain and weather that could assist or enable electronic warfare operations from both a friendly and threat viewpoint.
  o Maintain updated weather and terrain data.
  o Locate terrain for communications and non-communications sites, line of sight.
  o Identify aspects of terrain and weather that may have an impact on the electromagnetic spectrum.
- Define the threat.
  o Communications systems, including threat radio nets and network nodes.
  o Noncommunications emitters.
  o Electronic support systems.
  o Electronic attack systems.
- Identify host-nation use of the electromagnetic spectrum (restricted frequencies such as government, industry, and emergency responders).
- Identify friendly capabilities, shortfalls and readiness.
  o Electronic attack capabilities and status (joint and Army).
  o Location and availability of organic friendly electronic warfare capabilities (such as Prophets and counter-radio-controlled IED EW systems.
  o Electronic warfare vulnerabilities.
  o Equipment updates, both hardware and software.
- Identify enemy capabilities, shortfalls and readiness.
  o Electronic warfare capabilities and status (if known).
  o Electronic warfare vulnerabilities.
  o Electronic order of battle.

Electronic Warfare Target Folder
- Describe the targeted capability, its associated vulnerabilities, and the friendly capabilities used to engage them.
- Maintain updated high-value target and high-payoff target lists.
- Develop a prioritized target list based on high-value targets and high-payoff targets.
- EW target folders are split between traditional and asymmetric targets.
  o Traditional targets might include integrated air defense systems, communications nodes, and radar facilities. Traditional targets are normally fixed or less mobile than asymmetric targets and are easier to develop.
  o Asymmetric targets might include individual cell phones, radio-controlled improvised explosive devices, global positioning systems and wireless networks. Asymmetric targets can be either stationary or mobile and are typically harder to develop than traditional targets during the targeting phase.

Figure C-2. Sample update information to the electronic warfare running estimate

Appendix D
Electronic Warfare-Related Reports and Messages

This appendix provides information and references for electronic warfare and electronic warfare-related reports and message formats.

MESSAGES AND SUMMARIES

D-1. The following messages and summaries are associated with the planning, synchronization, deconfliction, and assessment of EW operations.

ELECTRONIC ATTACK DATA MESSAGE

D-2. An electronic attack data message reports an electronic attack strobe from an affected or detecting unit's position to an aircraft emitting an electronic attack. It is used to determine the location of a hostile or unknown aircraft emitting an electronic attack. The detecting unit reports its detection to all units using a given network when the data link is degraded or not operational.

D-3. Upon receipt of several messages, the source of enemy electronic attack can be determined by comparing lines of bearing from the different origins (triangulation).

D-4. See FM 6-99.2, page 83, for the format.

ELECTRONIC ATTACK REQUEST FORMAT

D-5. Electronic fires fall within three categories: preplanned, preplanned on-call, and immediate. Requesting airborne electronic attack support for ground operations is similar to requesting close air support. Requests for an electronic attack are sent via the normal joint air request process. Requesters use either a joint tactical air strike request or joint tactical air support request. (See FM 3-09.32 for a sample.) A theater-specific electronic attack request format may complement a joint tactical air strike request.

D-6. When submitting the request, the following information must be provided in the remarks section (section 8):

- Target location.
- Prioritized target description and jam frequencies.
- Time on target (window).
- Joint terminal attack controller.
- Jamming control authority call sign and frequency.
- Friendly force disposition (for example, troop movement route).
- Friendly frequency restrictions.
- Remarks.

ELECTRONIC WARFARE FREQUENCY DECONFLICTION MESSAGE

D-7. An EW frequency deconfliction message promulgates a list of protected, guarded, and taboo frequencies. This list allows friendly forces to use the frequency spectrum without adverse impact from friendly electronic attack. (See FM 6-99.2, page 86, for the format.)

Appendix D

ELECTRONIC WARFARE MISSION SUMMARY

D-8. The EW mission summary summarizes significant EW missions and reports the status of offensive EW assets. EW and electronic-attack-capable surface and air units use it to provide information on EW operations. Service components use it to report significant events for subsequent analysis. (See FM 6-99.2, page 87, for the format.)

ELECTRONIC WARFARE REQUESTING TASKING MESSAGE

D-9. Joint task force commanders use the electronic warfare requesting tasking message to task component commanders to perform EW operations in support of the joint EW plan and to support component EW operations. Component commanders use this message to request EW support from sources outside their command.

JOINT TACTICAL AIR STRIKE REQUEST OR JOINT TACTICAL AIR SUPPORT REQUEST

D-10. Use a joint tactical air strike request or joint tactical air support request to request electronic attack. These requests require the information listed in paragraph D-6. Organizations without an automated capability submit these requests using DD Form 1972 (Joint Tactical Air Strike Request). See JP 3-09.3 and FM 3-09.32 for more information.

JOINT SPECTRUM INTERFERENCE RESOLUTION

D-11. The joint spectrum interference resolution program replaced the DOD meaconing, intrusion, jamming, and interference program in June, 1992. Follow guidance in CJCSI 3320.02C to report incidents of spectrum interference.

JOINT RESTRICTED FREQUENCY LIST

D-12. Operational, intelligence, and support elements use the joint restricted frequency list to identify the level of protection desired for various networks and frequencies. The list should be limited to the minimum number of frequencies necessary for friendly forces to accomplish objectives.

D-13. See Annex A to appendix B to JP 3-13.1 for the joint restricted frequency list format. The format is used by the joint automated communications-electronics operations instruction system. The format is unclassified but should show the proper classification of each paragraph when filled in. (See CJCSI 3320.01B and JP 3-13.1 for additional information.)

COUNTER-IMPROVISED-EXPLOSIVE-DEVICE ACTIVITIES

D-14. Certain reports and references are associated with counter-improvised-explosive-device activities. Most of these reports include information pertinent to counter-radio-controlled improvised-explosive-device EW activities. EW working groups have the responsibility to monitor these reports to assess planned counter-radio-controlled improvised-explosive-device EW operations and to support future operations. These reports typically use formats established in FM 6-99.2 modified to include improvised explosive device considerations and current operations. See GTA 90-10-046 for examples of reports and references applicable to counter-radio-controlled improvised-explosive-device EW operations.

Appendix E

Army and Joint Electronic Warfare Capabilities

This appendix provides information on Army and other Service electronic warfare capabilities. It is not an all-inclusive list. Due to the evolving nature of electronic warfare equipment and systems, this information is perishable and should be augmented, updated, and maintained by the unit electronic warfare officer.

ARMY

E-1. The Army is currently expanding its electronic warfare (EW) capability. It maintains several EW systems in its inventory. Currently, all units whose sole purpose is to conduct EW operations are assigned to 1st Information Operations Command. When requested, these capabilities are provided to combatant commands for employment at corps and lower echelons.

COUNTER-RADIO-CONTROLLED IMPROVISED-EXPLOSIVE-DEVICE EW SYSTEMS

E-2. Counter-radio-controlled improvised-explosive-device EW systems form a family of electronic attack systems. Army forces use these systems to prevent improvised explosive device detonation by radio frequency energy. The Army maintains both a mounted and dismounted counter-radio-controlled improvised-explosive-device EW capability to protect personnel and equipment. For a detailed description of these systems, see appendix F.

AIRCRAFT SURVIVABILITY EQUIPMENT

E-3. Aircraft survivability equipment aims to reduce aircraft vulnerability, thus allowing aircrews to accomplish their immediate mission and survive. Army aviation maintains a suite of aircraft survivability equipment that provides protection against electronic attack. This protection can include radio frequency warning and countermeasures systems, a common missile warning system, information requirement countermeasures systems, and laser detection and countermeasure systems. For a detailed description of aircraft survivability equipment EW-related systems, see appendix F.

INTELLIGENCE SYSTEMS

E-4. The intelligence community maintains many systems that provide data for use in EW operations. Signals intelligence systems provide most of this required data. These assets are dual use. Usually the data collected is categorized as signals intelligence. It is maintained within sensitive compartmented information channels and governed by the National Security Agency/Central Security Service. The data sometimes support EW or, more specifically, electronic warfare support. Paragraphs E-5 through E-7 illustrate some intelligence systems that (when tasked) can provide electronic warfare support data to support electronic attack and electronic protection actions. For a detailed description of other intelligence and EW-support-related systems, see appendix F.

Appendix E

Guardrail Common Sensor

E-5. The Guardrail common sensor is a corps-level airborne signals intelligence collection and location system. (See figure E-1.) It provides tactical commanders with near real-time targeting information. Key features include the following: integrated communications intelligence and electronic intelligence reporting, enhanced signal classification and recognition, near real-time direction finding, precision emitter location, and an advanced integrated aircraft cockpit. Preplanned product improvements include frequency extension, computer-assisted online sensor management, upgraded data links, and the capability to exploit a wider range of signals. The Guardrail common sensor shares technology with the ground-based common sensor, airborne reconnaissance-low, and other joint systems.

Figure E-1. Guardrail common sensor

Aerial Common Sensor

E-6. The aerial common sensor is the Army's programmed airborne intelligence, surveillance, and reconnaissance system. (See figure E-2.) It will replace the current RC-7 airborne reconnaissance-low and Guardrail common sensor programs. The aerial common sensor uses the operational and technical legacies of the airborne reconnaissance-low and Guardrail common sensor systems as well as some technological improvements. This sensor will then provide a single, effective, and supportable multiple-intelligence system for the Army. The aerial common sensor will include a full multiple-intelligence capability, including carrying signals intelligence payloads, electro-optic and infrared sensors, radar payloads, and hyperspectral sensors.

Figure E-2. Aerial common sensor (concept)

Prophet

E-7. The Prophet system is the division, brigade combat team, and armored cavalry regiment principal ground tactical signals intelligence and EW system. (See figure E-3.) Prophet systems will also be assigned to the technical collection battalion of battlefield surveillance brigades. Prophet detects, identifies, and locates enemy electronic emitters. It provides enhanced situational awareness and actionable 24-hour information within the unit's area of operations. Prophet consists of a vehicular signals intelligence receiver mounted on a high mobility multipurpose wheeled vehicle, plus a dismounted-Soldier-portable version. The dismounted Soldier portable version is used for airborne insertion or early entry to support rapid reaction contingency and antiterrorist operations. Future Prophet systems are planned to include an electronic attack capability.

Figure E-3. Prophet (vehicle-mounted)

MARINE CORPS

E-8. The Marine Corps has two types of EW units: radio battalions (often called RADBNs), and Marine tactical EW squadrons (referred to as VMAQs). Paragraphs E-9 through E-24 discuss the units' missions, their primary tasks, and capabilities currently being employed. (For further information on the Marine Corps EW units and systems, see MCWP 2-22.)

RADIO BATTALION

E-9. Radio battalions are the Marine Corps' tactical level ground-based EW units. During operations, teams from radio battalions are most often attached to the command element (or senior headquarters) of Marine expeditionary units. Each radio battalion has the following mission, tasks, and equipment.

Appendix E

Mission and Tasks

E-10. The mission of the radio battalion is to provide communications security monitoring, tactical signals intelligence, EW, and special intelligence communication support to the Marine air-ground task force (MAGTF). The radio battalion's tasks include—

- Executing interception; radio direction finding; recording and analysis of communications and noncommunications signals; and signals intelligence processing, analysis, production, and reporting.
- Conducting EW against enemy or adversary communications.
- Helping protect MAGTF communications from enemy exploitation by conducting communications security monitoring, analysis, and reporting on friendly force communications.
- Providing special intelligence communications support and cryptographic guard (personnel and terminal equipment) in support of the MAGTF command element. Normally, the communications unit supporting the MAGTF command element provides communications connectivity for special intelligence communications.
- Providing task-organized detachments to MAGTFs with designated signals intelligence, EW, special intelligence communication, and other required capabilities.
- Exercising technical control and direction over MAGTF signals intelligence and EW operations.
- Providing radio reconnaissance teams with specialized insertion and extraction capabilities (such as combat rubber raiding craft, fast rope, rappel, helocast, and static-line parachute) for specified signals intelligence and limited electronic attack support during advance force, preassault, or deep postassault operations.
- Coordinating technical signals intelligence requirements and exchanging technical information and material with national, combatant command, joint, and other signals intelligence units.
- Providing intermediate, third, and fourth echelon maintenance of the radio battalion's signals intelligence and EW equipment.

Equipment

E-11. The following illustrate EW capabilities a radio battalion uses to accomplish the mission and perform the tasks in support of the MAGTF:

AN/ULQ-19(V)2 Electronic Attack Set

E-12. The AN/ULQ-19(V)2 electronic attack set allows operators to conduct spot or sweep jamming of single-channel voice or data signals. To provide the required jamming, the system must be employed and operated from a location with an unobstructed signal line of sight to the target enemy's communications transceiver.

AN/MLQ-36 Mobile Electronic Warfare Support System

E-13. The AN/MLQ-36 mobile electronic warfare support system provides a multifunctional capability that gives signals intelligence and EW operators limited armor protection. This equipment can provide signals intelligence and EW support to highly mobile mechanized and military operations in urban terrain where maneuver or armor protection is critical. This system is installed in a logistic variant of the Marine Corps's light armored vehicle. It consists of the following:

- Signals intercept system.
- Radio direction finding system.
- Electronic attack system.
- Secure communication system.
- Intercom system.

Army and Joint Electronic Warfare Capabilities

AN/MLQ-36A Mobile Electronic Warfare Support System (Product Improved)

E-14. The product-improved AN/MLQ-36A mobile electronic warfare support system (sometimes called the AN/MLQ-36A MEWSS PIP) is an advanced signals intelligence and EW system integrated into the Marine Corps's light armored vehicle. (See figure E-4.) This system replaces the equipment in the AN/MLQ-36.

E-15. The AN/AMLQ-36A has the following capabilities:
- Detect and evaluate enemy communications emissions.
- Detect and categorize enemy noncommunications emissions (such as battlefield radars).
- Determine lines of bearing.
- Degrade enemy tactical radio communications.

When mission-configured and working cooperatively with other AN/MLQ-36As, the system can provide precision location of battlefield emitters.

E-16. This system and its future enhancements will provide the capability to exploit new and sophisticated enemy electronic emissions and conduct electronic attack in support of existing and planned national, combatant command, fleet, and MAGTF signals intelligence and EW operations.

Figure E-4. AN/MLQ-36A mobile electronic warfare support system

MARINE TACTICAL ELECTRONIC WARFARE SQUADRON

E-17. Marine tactical electronic warfare squadrons are the Marine Corps's airborne tactical EW units. Each squadron has the following mission, tasks, and capabilities.

Mission and Tasks

E-18. The mission of the electronic warfare squadron is to provide EW support to the MAGTF and other designated forces. The squadron conducts tactical jamming to prevent, delay, or disrupt the enemy's ability to use the following kinds of radars: early warning, acquisition, fire or missile control, counterfire, and battlefield surveillance. Tactical jamming also denies and degrades enemy communication capabilities. The squadron conducts electronic surveillance operations to maintain electronic orders of battle. These include both selected emitter parameters and nonfriendly emitter locations. The squadron also provides threat warnings for friendly aircraft, ships, and ground units. Squadron tasks include—
- Providing airborne electronic attack and EW support to the aviation combat element and other designated operations by intercepting, recording, and jamming threat communications and noncommunications emitters.
- Processing, analyzing, and producing routine and time-sensitive electronic intelligence reports for updating and maintaining enemy electronic order of battle.

Appendix E

- Providing liaison personnel to higher staffs to assist in squadron employment planning.
- Providing an air EW liaison officer to the MAGTF EW coordination cell.
- Conducting electronic attack operations for electronic protection training of MAGTF units.

E-19. The squadron's EW division supports EA-6B Prowler tactical missions with intelligence, the tactical electronic reconnaissance processing and evaluation system (TERPES), and the joint mission planning system. All systems support premission planning and postmission processing of collected data, and production of pertinent intelligence reports. Working with squadron intelligence, these systems provide required electronic intelligence and electronic order of battle intelligence products to the aviation combat element, MAGTF, and other requesting agencies.

Equipment

E-20. Marine tactical electronic warfare squadrons maintain the following equipment:
- EA-6B Prowler.
- Joint mission planning system.
- Tactical electronic reconnaissance processing and evaluation system.

EA-6B Prowler

E-21. The EA-6B Prowler is a subsonic, all-weather, carrier-capable aircraft. (See figure E-5.) The crew consists of one pilot and three electronic countermeasure officers. The EA-6B has two primary missions. One is collecting and processing designated threat signals of interest for jamming and subsequent processing, analysis, and intelligence reporting. The other is employing the AGM-88 high-speed antiradiation missile against designated targets. The EA-6B's AN/ALQ-99 tactical jamming system incorporates receivers for the reception of emitted signals and external jamming pods for the transmission of energy to jam targeted radars (principally those associated with enemy air defense radars and associated command and control). In addition to the AN/ALQ-99, the EA-6B also employs the USQ-113 communications jammer to collect, record, and disrupt threat communications.

Figure E-5. EA-6B Prowler

Army and Joint Electronic Warfare Capabilities

Joint Mission Planning System

E-22. The joint mission planning system helps the EA-6B aircrew plan and optimize receivers, jammers, and high-speed antiradiation missiles. This system allows an operator to—
- Maintain area of operations emitter listings.
- Edit emitter parameters.
- Develop mission-specific geographic data and electronic order of battle to—
  - Tailor or create high-speed antiradiation missile direct attack libraries, or manually modify entries or new threat cards.
  - Plan target selection.
- Perform postflight mission analysis to—
  - Identify electronic emitters using various electronic parameter databases and electronic intelligence analytical techniques.
  - Localize emitters by coordinates with a certain circular error of probability for each site.
  - Correlate new information with existing data.
  - Gather postflight high-speed antiradiation missile information. This information includes aircraft launch parameters, predicted seeker footprint, and the onboard system detection of a targeted signal at impact.

AN/TSQ-90 Tactical Electronic Reconnaissance Processing and Evaluation System

E-23. The TERPES (AN/TSQ-90) is an air and land transportable, single-shelter electronic intelligence processing and correlation system. Each of the four Marine tactical electronic warfare squadrons includes a TERPES section.

E-24. A TERPES section consists of Marines, equipment, and software. The section identifies and locates enemy radar emitters from data collected by EA-6B aircraft and those received from other intelligence sources. It processes and disseminates EW data rapidly to MAGTF and other intelligence centers and provides mission planning and briefing support. Section support areas include operational support, intelligence analysis support, data fusion, fusion processing, and intelligence reporting. The section provides the following operational support:
- Translates machine-readable, airborne-collected, digital data into human- and machine-readable reports (such as paper, magnetic tape, secure voice, plots, and overlays).
- Receives and processes EA-6B mission tapes.
- Accepts, correlates, and identifies electronic emitter data from semiautomatic or automatic collection systems using various electronic parameter databases and various analysis techniques.
- Provides tactical jamming analysis.

Appendix E

AIR FORCE

E-25. The Air Force has two primary platforms that provide EW capability: the EC-130H Compass Call and RC-135V/W Rivet Joint. (For further information on Air Force EW equipment, see AFDD 2-5.1.)

EC-130H COMPASS CALL

E-26. The EC-130H Compass Call is an airborne tactical weapon system. (See figure E-6.) Paragraphs E-27 through E-31 discuss the EC-130H missions, primary tasks, and capabilities.

Mission and Tasks

E-27. The EC-130H's mission is to disrupt enemy command and control information systems and limit the coordination essential for force management. The EC-130H's primary task is to employ offensive counterinformation and electronic attack capabilities in support of U.S. and multinational tactical air, surface, and special operations forces.

Figure E-6. EC-130H Compass Call

Capabilities

E-28. The EC-130H is designed to deny, degrade, and disrupt adversary command and control information systems. This includes denial and disruption of enemy surveillance radars; denial and disruption of hostile communications being used in support of enemy ground, air, or maritime operations; and denial and disruption of many modern commercial communication signals that an adversary might employ.

Army and Joint Electronic Warfare Capabilities

> **Compass Call During Operation Iraqi Freedom**
>
> During Operation Iraqi Freedom, much speculation appeared in the press about why Iraqi forces failed to ignite the oil facilities they had wired for destruction. During the coalition's seizure of Al Faw, Compass Call disrupted the Iraqi regime's control of its troops by jamming its communications. Instead of receiving orders to detonate the oil terminals, Iraqi troops heard only the ratcheting static of Compass Call jamming until coalition ground troops had secured the area. In addition to the conquest of the Al Faw Peninsula, successful military operations supported by Compass Call in Operation Iraqi Freedom included the seizure of four airfields; two successful prisoner of war rescues; and the ground offensive from Basrah to Nasariyah, Najaf, Baghdad, and Tikrit. In all these instances, Compass Call jamming prevented a trained, experienced enemy from coordinating actions against coalition forces.
>
> "EC-130H Compass Call: A textbook example of Joint Force integration at its best", Electronic Warfare Working Group, U.S. House of Representatives, Issue Brief #17, 11 Mar 2004. (Available at http://www.house.gov/pitts/initiatives/ew/Library/Briefs/brief17.htm)

RC-135V/W RIVET JOINT

E-29. Paragraphs E-30 through E-31 discuss the missions, primary tasks, and capabilities of the RC-135V platforms.

Mission and Tasks

E-30. The RC-135V/W Rivet Joint is a combatant-command-level surveillance asset that responds to national-level taskings. (See figure E-7.) Its mission is to support national consumers, combatant commanders, and combat forces with direct, near real-time reconnaissance information and electronic warfare support. It collects, analyzes, reports, and exploits information from enemy command and control information systems. During most contingencies, it deploys to the theater of operations with the airborne elements of the theater air control system.

Figure E-7. RC-135V/W Rivet Joint

Capabilities

E-31. The RC-135V/W is equipped with an extensive array of sophisticated intelligence gathering equipment that enables monitoring of enemy electronic activity. The aircraft is integrated into the theater air control system via data links and voice (as required). Refined intelligence data can be transferred from Rivet Joint to an Airborne Warning and Control System platform through the tactical digital information link. Alternatively, this data can be placed into intelligence channels via satellite and the tactical information broadcast service (a near real-time combatant command information broadcast). The aircraft

Appendix E

has secure ultrahigh frequency, very high frequency, and high frequency (commonly known as UHF, VHF, and HF respectively) as well as satellite communications. It can be refueled in the air.

NAVY

E-32. The Navy's primary airborne EW platforms are the EA-6B Prowler and its planned replacement, the E/A-18G Growler. E/A-18G fielding is scheduled to begin in 2009 and is scheduled to replace the Navy's carrierborne EA-6B aircraft. The Navy also maintains both surface and subsurface EW shipboard systems for offensive and defensive missions in support of the fleet. (For further information on Navy missions and equipment, see NWP 3-13.)

EA-6B PROWLER

E-33. Paragraphs E-34 through E-39 discuss the missions, primary tasks, and capabilities of the Navy's EA-6B Prowler platforms. (See figure E-8.)

Figure E-8. Navy EA-6B Prowler

Mission and Tasks

E-34. The mission of the Navy's EA-6B Prowler is to ensure survivability of U.S. and multinational forces through suppression of enemy air defenses (using the radar-jamming AN/ALQ-99 tactical jamming system), lethal suppression (using the AGM-88 high-speed antiradiation missile), and communications jamming (using the USQ-113 radio countermeasures set). Prowlers have supported U.S. and multinational forces operating from various expeditionary sites throughout the world while maintaining full presence on all Navy aircraft carriers.

Capabilities

E-35. The Navy's EA-6B Prowlers are outfitted with either the improved capability II or improved capability III systems. The following lists the major capability upgrades these systems provide.

Improved Capability II

E-36. The improved capability II program was initiated in the 1980s. It was completed across the fleet of EA-6B aircraft (including U.S. Marine Corps aircraft) in the 1990s. The program incorporated incremental capability improvements that include communications, navigation, and computer interface upgrades; a high-speed antiradiation missile capability; and improved jamming pods. Several system interfaces were also upgraded in preparation for the improved capability III improvements.

Army and Joint Electronic Warfare Capabilities

Improved Capability III

E-37. The improved capability III program incorporates a highly evolved receiver system and provides upgraded EA-6B aircraft with increased signal detection, geolocation capability, a new selective reactive-jamming capability, and better reliability. High-speed antiradiation missile employment is also improved due to the speed of the receiver and its geolocation accuracy. Increased battlefield situational awareness of joint forces is also provided through Link-16. The improved capability III program provides a new ALQ-218 receiver system, integration of the USQ-113 and the multifunctional information distribution system (often called MIDS). This system incorporates Link-16 and various connectivity avionics into the Prowler. The major EW-related subsystems are the AN/ALQ-99 (V) tactical jamming countermeasures set and AN/USQ-113 (V) radio countermeasures set.

E-38. The AN/ALQ-99 (V) tactical jamming countermeasures set has upgraded receivers and processors to provide the following:

- Improved frequency coverage.
- Direction-of-arrival determination capability.
- Narrower frequency discrimination to support narrowband jamming.
- Enhanced interface with onboard systems.

E-39. The AN/USQ-113 (V) radio countermeasures set will enhance the aircraft's jamming capability through its integration with the tactical display system. This will enable the crew to display AN/USQ-113 communications jamming data as well as control AN/USQ-113 operations through the tactical display system.

E/A-18G GROWLER

E-40. The E/A-18G Growler is the Navy's replacement aircraft for the EA-6B Prowler. Paragraphs E-41 and E-42 discuss the missions, primary tasks, and capabilities of the Navy's E/A-18G Growler. (See figure E-9.) E/A-18G fielding began in 2008. The first operational E/A-18G deployment will occur in 2009, as the Navy begins to replace its carrierborne EA-6B aircraft.

Figure E-9. EA-18 Growler

Mission and Tasks

E-41. The EA-18G can detect, identify, locate, and suppress hostile emitters. It will provide enhanced connectivity to national, combatant command, and strike assets. Additionally, the EA-18G will provide organic accurate emitter targeting using on-board suppression weapons, such as the high-speed antiradiation missile.

Appendix E

Capabilities

E-42. The following is a list of the E/A-18G's general capabilities:
- Suppression of enemy air defenses. The EA-18G will counter enemy air defenses using both reactive and preemptive jamming techniques.
- Stand-off and escort jamming. The EA-18G will be highly effective in the traditional stand-off jamming mission, but with the speed and agility of a Super Hornet, it will also be effective in the escort role.
- Integrated air and ground airborne electronic attack. Enhanced situational awareness and uninterrupted communications will enable the EA-18G to achieve a higher degree of integration with ground operations than previously.
- Self-protect and time-critical strike support. With its active electronically scanned array radar, digital data links, and air-to-air missiles, the EA-18G will be able to protect itself and effectively identify and prosecute targets.
- Growth. High commonality with the F/A-18E and F/A-18F, nine available weapon stations, and modern avionics enable cost-effective synergistic growth, setting the stage for continuous capability enhancement.

E-43. The following is a list of the E/A-18G's airborne electronic attack capabilities:
- Entire spectrum. The EA-18G's ALQ-218 wideband receiver combined with the ALQ-99 tactical jamming system will be effective against any surface-to-air threat.
- Precision airborne electronic attack. Selective-reactive technology enables the EA-18G to rapidly sense and locate threats much more accurately than before. This improved accuracy enables greater concentration of energy against threats.
- Advanced communication countermeasures. Its modular communication countermeasure set enables the EA-18G to counter a wide range of communication systems and is readily adaptable to an ever changing threat spectrum.
- Interference cancellation system. This system dramatically enhances aircrew situational awareness by enabling uninterrupted communications during jamming operations.

CAPABILITIES SUMMARY

E-44. Table E-1 lists Army and joint EW capabilities. (Bold text indicates capabilities not described in the preceding paragraphs.) EW officers, noncommissioned officers, and supporting staff members should be familiar with these capabilities and how they can support Army operations. Additional information on the EW capabilities listed in table E-1 is found in the Web sites listed in table E-2, page E-12.

Table E-1. Army and joint electronic warfare capabilities

	Army	Air Force	Navy	Marine Corps
Airborne	RC-12 Guardrail	EC-130J Commando Solo	EA-6B Prowler	EA-6B Prowler
	airborne common sensor	EC-130H Compass Call	EA-18G Growler	
		RC-135V/W Rivet Joint	EP-3E Aries II	
		F-16CJ		
		E-8 JSTARS		
Unmanned aircraft system*	RQ-5A/MQ-5B Hunter (Corps)	RQ-4A (Joint) Global Hawk	RQ-2 Pioneer	
	RQ-7A/B Shadow (brigade)	RQ-1L (Joint) Predator	MQ-8B Fire Scout Vertical Take-off	
	MQ-1C/Sky Warrior (replacement for Hunter)	RQ-11 Raven	Silver Fox	
	MQ-8 Fire Scout			RQ-11 Raven
	RQ-11 Raven (battalion) Hand Launched			Scan Eagle
Ground	AN/MLQ-40 Prophet		CREW Systems (Joint)	AN/MLQ-36 MEWSS
Note:	*Other Services may refer to unmanned aircraft systems as unmanned aerial systems or vehicles.			
CREW	counter radio-controlled improvised explosive device electronic warfare			
MEWSS	mobile electronic warfare support system			
SOF	special operations forces			

Appendix E

Table E-2. Electronic warfare systems and platforms resources

Army platforms and systems http://www.sed.monmouth.army.mil/avionics/ http://www.sec.army.mil/secweb/fact_sheets/fact_sheets.php
Air Force platforms and systems http://www.af.mil/factsheets/factsheet.asp?fsID=182 http://www.airforce-technology.com/projects/#Unmanned_Aerial_Vehicles_(UAV_/_UCAV)
Navy systems platforms and systems http://acquisition.navy.mil/programs http://www.naval-technology.com/projects/ http://www.navy.mil/navydata/fact.asp
Marine Corps platforms and systems http://www.marcorsyscom.usmc.mil/sites/cins/INTEL/USMC%20CREW/index.html
Joint programs https://www.jieddo.dod.mil

Appendix F
Tools and Resources Related to Electronic Warfare

This appendix provides information on tools and reachback resources related to electronic warfare. Electronic warfare officers, noncommissioned officers, and supporting staff members should be familiar with these tools and resources and how to use them to support electronic warfare operations. Some tools and resources require an approved user account prior to being granted access.

ARMY REPROGRAMMING ANALYSIS TEAM

F-1. The Army Reprogramming Analysis Team (ARAT) supports tactical commanders. It provides timely reprogramming of any Army-supported software used for target acquisition, target engagement, measurement and signature intelligence, and vehicle and aircraft survivability (including that operated by other Services). The team provides software changes not readily possible by operator input to respond to rapid deployments or changes in the operational environment. See their Web site at https://ako.sec.army.mil/arat/index.html (Army Knowledge Online login required).

F-2. ARAT provides reprogramming support to counter-radio-controlled improvised-explosive-device (IED) electronic warfare (EW) (sometimes referred to as CREW), and other electronic systems.

F-3. The team is accessible via the Army Reprogramming Analysis Team's Warfighter Survivability Software Support Portal. A secure Internet protocol router network (SIPRNET) account is required to access the portal.

NATIONAL GROUND INTELLIGENCE CENTER

F-4. The National Ground Intelligence Center provides all-source analysis of the threat posed by IEDs produced and used by foreign terrorist and insurgent groups. The center supports U.S. forces during training, operational planning, deployment, and redeployment.

F-5. The center maintains a counter-IED targeting program (often called CITP) portal on its SIPRNET site. This portal provides information concerning IED activities and incidents as well as IED assessments.

ELECTRONIC ORDER OF BATTLE

F-6. An electronic order of battle details all known combinations of emitters and platforms in a particular area of responsibility. It consists of several reachback resources:
- National Security Agency-Electronic Intelligence Parameter Query.
- U.S. electromagnetic systems database.
- National Ground Intelligence System parametric information relational intelligence tool database.
- Military equipment parametrics and engineering database.

E-SPACE

F-7. E-Space is a Department of Defense (DOD) entity housed in the National Security Agency. It provides intelligence assistance (primarily signals intelligence) to deployed EW officers. E-Space is a reachback capability available to EW officers and spectrum managers that can be leveraged to provide all-source intelligence products and answers to requests for information and spectrum interference questions.

Appendix F

JOINT ELECTRONIC WARFARE CENTER

F-8. The Joint Electronic Warfare Center is DOD's only joint EW center of expertise. It provides EW subject matter expertise from a range of backgrounds, including people with current multi-Service operational experience. The center has a limited capability to perform modeling and simulation studies and EW red team support. It can deploy in a support role if approved by the U.S. Strategic Command.

JOINT IMPROVISED EXPLOSIVE DEVICE DEFEAT ORGANIZATION

F-9. The Joint Improvised Explosive Device Defeat Organization (known as JIEDDO) leads, advocates, and coordinates all DOD actions in support of efforts by combatant commanders and their joint task forces to defeat IEDs as weapon of strategic influence.

JOINT SPECTRUM CENTER

F-10. The Joint Spectrum Center ensures DOD effectively uses the electromagnetic spectrum in support of national security and military objectives. The center serves as DOD's center of excellence for electromagnetic spectrum management matters in support of the combatant commands, military departments, and DOD agencies in planning, acquisition, training, and operations.

F-11. The center maintains databases and provides data about friendly force command and control information system locational and technical characteristics. This information is used to plan electronic protection measures. These databases provide EW planners with information covering communication, radar, navigation, broadcast, identification, and EW systems operated by the DOD, other government agencies, and private businesses and organizations.

F-12. The center provides information on a quick-reaction basis in various formats and media to support EW planners and spectrum managers.

KNOWLEDGE AND INFORMATION FUSION EXCHANGE

F-13. The Knowledge and Information Fusion Exchange (sometimes called KnIFE) is a program sponsored by U.S. Joint Forces Command. It provides Soldiers with observations, insights, and lessons from operations around the world.

ADDITIONAL INFORMATION

F-14. Further information on the above tools and resources can be accessed through Army Knowledge Online. The links to these Web sites can be viewed by first accessing the "Army Operational Electronic Warfare Course" on Army Knowledge Online at http://www.us.army.mil/suite/page/400055 and then clicking on Folders >Links>EW links.

Glossary

SECTION I – ACRONYMS AND ABBREVIATIONS

ARAT	Army Reprogramming Analysis Team
C-3	operations directorate of a multinational (combined) staff
CJCSI	Chairman of the Joint Chiefs of Staff instruction
CJCSM	Chairman of the Joint Chiefs of Staff manual
COA	course of action
DD	Department of Defense (official forms only)
DOD	Department of Defense
DODI	Department of Defense Instruction
EW	electronic warfare
FM	field manual
FMI	field manual, interim
G-2	assistant chief of staff, intelligence
G-3	assistant chief of staff, operations
G-5	assistant chief of staff, plans
G-6	assistant chief of staff, signal
G-7	assistant chief of staff, information engagement
GTA	graphic training aid
HF	high frequency
Hz	hertz
IED	improvised explosive device
IO	information operations
IPB	intelligence preparation of the battlefield
ISR	intelligence, surveillance, and reconnaissance
J-2	intelligence directorate of a joint staff
J-3	operations directorate of a joint staff
J-5	plans directorate of a joint staff
J-6	communications system directorate of a joint staff
JFMO	Joint Frequency Management Office
JIOWC	Joint Information Operations Warfare Center
JP	joint publication
MAGTF	Marine air-ground task force
MC	Military Committee (NATO)
MCWP	Marine Corps warfighting publication
MDMP	military decisionmaking process

Glossary

NATO	North Atlantic Treaty Organization
S-2	intelligence staff officer
S-3	operations staff officer
S-6	signal staff officer
S-7	information engagement staff officer
SIPRNET	SECRET Internet Protocol Router Network
STANAG	standardization agreement (NATO)
TERPES	tactical electronic reconnaissance processing and evaluation system
U.S.	United States

SECTION II – TERMS

communications security
  (joint) The protection resulting from all measures designed to deny unauthorized persons information of value that might be derived from the possession and study of telecommunications, or to mislead unauthorized persons in their interpretation of the results of such possession and study. (JP 6-0)

computer network operations
  (joint) Comprised of computer network attack, computer network defense, and related computer network exploitation enabling operations. (JP 3-13)

directed energy
  (joint) An umbrella term covering technologies that relate to the production of a beam of concentrated electromagnetic energy or atomic or subatomic particles. (JP 3-13.1)

electromagnetic environment
  (joint) The resulting product of the power and time distribution, in various frequency ranges, of the radiated or conducted electromagnetic emission levels that may be encountered by a military force, system, or platform when performing its assigned mission in its intended operational environment. It is the sum of the electromagnetic interference; electromagnetic pulse; hazards of electromagnetic radiation to personnel, ordnance, and volatile materials; and natural phenomena effects of lightning and precipitation static. (JP 3-13.1)

electromagnetic environmental effects
  The impact of the electromagnetic environment upon the operational capability of military forces, equipment, systems, and platforms. It encompasses all electromagnetic disciplines, including electromagnetic compatibility and electromagnetic interference; electromagnetic vulnerability; electromagnetic pulse; electronic protection, hazards of electromagnetic radiation to personnel, ordnance, and volatile materials; and natural phenomena effects of lightning and precipitation static. (JP 3-13.1)

electromagnetic spectrum
  (joint) The range of frequencies of electromagnetic radiation from zero to infinity. It is divided into 26 alphabetically designated bands. (JP 1-02)

electromagnetic vulnerability
  (joint) The characteristics of a system that cause it to suffer a definite degradation (incapability to perform the designated mission) as a result of having been subjected to a certain level of electromagnetic environmental effects. (JP 1-02)

Glossary

electronic attack

(joint) Division of electronic warfare involving the use of electromagnetic energy, directed energy, or antiradiation weapons to attack personnel, facilities, or equipment with the intent of degrading, neutralizing, or destroying enemy combat capability and is considered a form of fires. (JP 3-13.1)

electronic protection

(joint) Division of electronic warfare involving actions taken to protect personnel, facilities, and equipment from any effects of friendly or enemy use of the electromagnetic spectrum that degrade, neutralize or destroy friendly combat capability. (JP 3-13.1)

electronic warfare

(joint) Military action involving the use of electromagnetic and directed energy to control the electromagnetic spectrum or to attack the enemy. Electronic warfare consists of three divisions: electronic attack, electronic protection, and electronic warfare support. (JP 3-13.1)

electronic warfare support

(joint) Division of electronic warfare involving actions tasked by, or under direct control of, an operational commander to search for, intercept, identify, and locate or localize sources of intentional and unintentional radiated electromagnetic energy for the purpose of immediate threat recognition, targeting, planning, and conduct of future operations. (JP 3-13.1)

emission control

(joint) The selective and controlled use of electromagnetic, acoustic, or other emitters to optimize command and control capabilities while minimizing, for operations security: a. detection by enemy sensors; b. mutual interference among friendly systems; and/or c. enemy interference with the ability to execute a military deception plan. (JP 1-02)

joint restricted frequency list

(joint) A time a geographically-oriented listing of TABOO, PROTECTED, and GUARDED functions, nets, and frequencies. It should be limited to the minimum number of frequencies necessary for friendly forces to accomplish objectives. (JP 3-13.1)

working group

(Army) A temporary grouping of predetermined staff representatives who meet to coordinate and provide recommendations for a particular purpose or function. (FMI 5-0.1)

References

REQUIRED PUBLICATIONS
These documents must be available to intended users of this publication.

FM 1-02 (101-5-1). *Operational Terms and Graphics.* 21 September 2004.
JP 1-02. *Department of Defense Dictionary of Military and Associated Terms.* 12 April 2001. (As amended through 4 March 2008.)
JP 3-13.1. *Electronic Warfare.* 25 January 2007.
FM 3-0. *Operations.* 27 February 2008.
FM 5-0 (101-5). *Army Planning and Orders Production.* 20 January 2005.
FM 6-0. *Mission Command: Command and Control of Army Forces.* 11 August 2003.
FMI 5-0.1. *The Operations Process.* 31 March 2006.

RELATED PUBLICATIONS
These documents contain relevant supplemental information.

JOINT AND DEPARTMENT OF DEFENSE PUBLICATIONS
Most joint publications are available online: <http://www.dtic.mil/doctrine/jpcapstonepubs.htm.>

CJCSI 3320.01B *Electromagnetic Spectrum Use in Joint Military Operations.* 01 May 2005
CJCSI 3320.02C. Joint Spectrum Interference Resolution (JSIR). 27 January 2006 (with change 1 as of 25 February 2008).
CJCSI 3320.03A *Joint Communications Electronics Operation Instructions.* 11 June 2005.
CJCSM 3122.03C. *Joint Operation Planning and Execution System Volume II, Planning Formats and Guidance.* 17 August 2007.
CJCSM 3320.01B *Joint Operations in the Electromagnetic Battlespace.* 25 March 2006.
CJCSM 3320.02A *Joint Spectrum Interference Resolution (JSIR) Procedures.* 16 February 2006.
DODI 4650.01. Policy and Procedures for Management and Use of the Electromagnetic Spectrum. 09 January 2009.
JP 2-0. *Joint Intelligence.* 22 June 2007.
JP 2-01. *Joint and National Intelligence Support to Military Operations.* 07 October 2004.
JP 3-0. *Joint Operations.* 17 September 2006.
JP 3-09. *Joint Fire Support.* 13 November 2006.
JP 3-09.3. *Joint Tactics, Techniques, and Procedures for Close Air Support (CAS).* 03 September 2003.
JP 3-13. *Information Operations.* 13 February 2006.
JP 3-13.3. *Operations Security.* 29 June 2006.
JP 3-13.4 (JP 3-58). *Military Deception.* 13 July 2006.
JP 3-60. *Joint Targeting.* 13 April 2007.
JP 6-0. *Joint Communications System.* 20 March 2006.

References

ARMY PUBLICATIONS
Most Army doctrinal publications are available online:
<http://www.army.mil/usapa/doctrine/Active_FM.html>.

FM 2-0 (34-1). *Intelligence.* 17 May 2004.

FM 3-09.32. JFIRE: *Multi-Service Tactics, Techniques, and Procedures for the Joint Application of Firepower.* 20 December 2007.

FM 3-13 (100-6). *Information Operations: Doctrine, Tactics, Techniques, and Procedures.* 28 November 2003.

FM 3-13.10 (3-51.1). *Multi-Service Tactics, Techniques, and Procedures for the Reprogramming of Electronic Warfare and Target Sensing Systems.* 22 January 2007.

FM 5-19 (100-14). *Composite Risk Management.* 21 August 2006.

FM 6-20-10. *Tactics, Techniques, and Procedures for the Targeting Process.* 8 May 1996.

FM 6-99.2 (101-5-2). *U.S. Army Report and Message Formats.* 30 April 2007.

FM 34-130. *Intelligence Preparation of the Battlefield.* 8 July 1994.

FMI 2-01. *Intelligence, Surveillance, and Reconnaissance (ISR) Synchronization.* 11 November 2008.

GTA 90-10-046. *MNC-I Counter IED Smart Book.* September 2008.

NATO PUBLICATIONS
Allied Joint Publication 3.6. *Allied Joint Electronic Warfare Doctrine.* December 2003.

MC 64. *NATO Electronic Warfare (EW) Policy.* 26 April 2004.

STANAG 5048 C3 (Edition 5). *The Minimum Scale of Convectivity for Communications and Information Systems for NATO Land Forces.* 16 February 2000.

OTHER PUBLICATIONS
AFDD 2-1.9. *Targeting.* 8 June 2006.

AFDD 2-5.1. *Electronic Warfare.* 5 November 2002.

Executive Order 12333. *United States Intelligence Activities.* 4 December 1981.

MCWP 2-22 (2-15.2). *Signals Intelligence.* 13 July 2004.

NWP 3-13. *Navy Information Operations.* June 2003.

SOURCES USED
Electronic Warfare Working Group, U.S. House of Representatives, Issue Brief #17. "Compass Call During Operation Iraqi Freedom." 11 March 2004. Available online at http://www.house.gov/pitts/initiatives/ew/Library/Briefs/brief17.htm.

PRESCRIBED FORMS
None

REFERENCED FORMS
DA Forms are available on the APD website (www.apd.army.mil. DD forms are available on the OSD website (www.dtic.mil/whs/directives/infomgt/forms/formsprogram.htm).

DA Form 2028. *Recommended Changes to Publications and Blank Forms.*

DD Form 1972. *Joint Tactical Air Strike Request.*

Index

Entries are by paragraph number unless specified otherwise.

A–B

aerial common sensor, E-6
aircraft survivability equipment, E-3
AN/ALQ-99 tactical jamming system, E-21, E-34
AN/MLQ-36 mobile electronic warfare support system, E-13, E-14
AN/MLQ-36A mobile electronic warfare support system, E-14–E-16
AN/TSQ-90 tactical electronic reconnaissance processing and evaluation system (TERPES), E-19, E-23–E-24
AN/ULQ-19(V)2 electronic attack set, E-12
antiradiation missiles, electronic attack and, 1-12, 1-54, 4-68. *See also* high-speed antiradiation missiles.
area denial, 1-11, 2-13, 2-16
  directed energy and, A-6
Army Reprogramming Analysis Team (ARAT), 5-16, F-1–F-3
assessment, defined, 4-79
  electronic attack and, 4-47–4-49
  electronic attack, of, 4-78, 4-79–4-83
asset management, 5-2, 5-13
asset tracking, EW support of, 2-14, 2-15
attack guidance matrix, electronic attack and, 4-44, 4-45, 4-46, 4-76
band designators, A-3, table A-1
battalion, EW working group at, 3-8
battle damage assessment, electronic attack and, 4-48, 4-64
battlefield coordination detachment, EW coordination and, 5-5
  EW support requests and, 4-78
battlefield surveillance brigade, Prophet and, E-7
Big Crow Program Office, 7-3
branches, EW supporting tasks for, 4-22, 4-23, 4-75
brigade, EW working group at, 3-6–3-7

C

center of gravity analysis, EW contributions to, 4-10–4-11, 4-39, figure 4-2
Central Security Service, 7-11, E-4
CITP, F-5
collateral damage, preventing, 1-54, 2-13, 4-21, 4-66
collection manager, 3-11
collection plan, 5-15
  electronic order of battle and, 3-13
  EW tasks in, 3-11
  preparation and, 4-76
  targeting and, 4-45
combat assessment, electronic attack and, 4-47–4-49
  execution, during, 4-83
command and control tasks, EW support to, table 2-2
command and control warfare, 2-7, table 2-1
  defined, 2-8
  electronic attack and, 4-45
  EW coordination with, 5-17
  EW support to, table 2-2
  EW synchronization and, 5-20
  fires warfighting function and, 2-13
  in a time-constrained environment, 4-33
  intelligence, surveillance, and reconnaissance and, 4-45
  preparation for, 4-76
command and control warfare working group, execution, during, 4-78
command and control warfighting function, EW support of, 2-15
commander's critical information requirements, course of action analysis and, 4-23
  course of action approval and, 4-28, 4-29
  execution, during, 4-78
commander's visualization, EW employment, of, 2-3
communications fratricide, 3-14
  EW synchronization and, 5-20
communications security, 4-69, 4-71, 7-5, E-10
company, EW support at, 3-9
Compass Call, E-26–E-28
composite risk management, assessment during, 4-81. *See also* EW risks, risk controls.
contingency planning, peacetime, joint, 5-3
continuing activities, EW contributions to, 4-34
convoy planning, EW support of, 2-14
coordination, EW, joint level, 5-3
  external EW agencies, with, 5-6
counter-IED targeting program (CITP), F-5
countermeasures, 1-24–1-26, A-7, E-3
  defined, 1-24
  degradation, 1-49
  electronic attack and, 1-9, 1-13
  protection warfighting function and, 2-16
  wartime reserve modes and, 1-43
counter-radio-controlled IED EW, 4-5, E-2
  deconfliction and, 5-18

25 February 2009   FM 3-36   Index-1

Index

Entries are by paragraph number unless specified otherwise.

counter-radio-controlled IED EW *(continued)*
 defensive electronic attack and, 1-13
 electronic protection and, 4-72
 EW reports and, D-14
 movement and maneuver warfighting function and, 2-11
 protection warfighting function and, 2-16
 reprogramming support to, F-2
 spectrum management and, 5-10
 sustainment warfighting function and, 2-14
 systems, E-2
course of action analysis, EW contributions to, 4-21–4-23
course of action approval, EW contributions to, 4-27–4-29
course of action comparison, EW contributions to, 4-24–4-26
course of action development, EW contributions to, 4-15–4-20
CREW. *See* counter-radio-controlled IED EW.
crisis action planning, joint, 5-3
critical vulnerabilities, identifying, figure 4-2
crowd control, directed energy and, A-6, A-7
cryptographic guard, radio battalion (Marine Corps) and, E-10
current operations cell, EW running estimate and, C-3

D

deception, 1-50, 2-16
 electronic, disruption and degradation and, 1-51
 EW support of, 4-18
decisionmaking in a time-constrained environment, EW working group decisionmaking tasks, 4-32–4-33
deconfliction, 5-18–5-19
 frequency, 6-11
 Joint Spectrum Center and, 7-8
 preparation and, 4-76
 protection and, 1-52
 spectrum requirements, 5-10–5-11
Defense Information Systems Agency, 7-4

Joint Spectrum Center and, 7-8
Defense Spectrum Organization, Joint Spectrum Center and, 7-8
defensive electronic attack, 4-61
degradation, 1-49, 1-51
denial, 1-49
destruction, 1-53
detection, 1-48
directed energy, A-6–A-8
 defined, 1-11
 doctrine development for, A-8
 electronic attack and, 1-9, 1-12
 EW and, A-8
 jamming, compared with, 4-68
directed-energy warfare, defined, A-6
directed-energy weapon, defined, A-6
disruption, 1-51

E

EA-6B Prowler, Marine Corps, E-19, E-21, E-24
 Navy, E-33–E-39
E/A-18G Growler, E-40–E-43
EC-130H Compass Call, E-26–E-28
electromagnetic compatibility, defined, 1-44
Electromagnetic Compatibility Center, 7-8
electromagnetic deception, control of, 3-11
 coordination of, 3-15
 defined, 1-27
 electronic attack and, 1-9, 1-10
electromagnetic effects, A-4
electromagnetic emissions, EW deconfliction and, 5-18
electromagnetic environment, described, A-1
 IPB and, 4-37
electromagnetic hardening, defined, 1-37
electromagnetic interference, defined, 1-38
 resolution of, 3-15
electromagnetic intrusion, defined, 1-28
electromagnetic jamming, defined, 1-29
electromagnetic pulse, defined, 1-30

electromagnetic spectrum, 1-47, figure A-1
 coordinating use of, 6-11
 defined, A-2
 EW deconfliction and, 5-18
 operations in, 1-4–1-7
 situational awareness, of, 2-16
electromagnetic spectrum management, 5-9–5-11
 defined, 1-42
electromagnetic vulnerability, defined, A-5
electronic attack, 1-9–1-13
 activities, 1-23–1-31
 AN/MLQ-36A capability for, E-16
 assessment of, 4-78
 battle damage assessment for, 4-64
 command and control warfare and, 4-45
 control of, 3-11
 coordination of, 4-62, 4-65
 counter-radio-controlled IED EW and, E-2
 deconfliction of, 4-62–4-63, 4-66
 defensive, 1-13, 1-14
 defensive and offensive compared, 4-61
 directed energy and, A-7
 disruption and degradation and, 1-51
 E/A-18G capabilities, E-42, E-43
 EC-130H capabilities, E-27–E-38
 electromagnetic spectrum, and the, 1-47
 electronic protection, compared, 1-14
 employment considerations, 4-61–4-68
 EW contributions to, 4-43
 EW deconfliction and, 5-18
 EW officer and, 3-12
 EW risks and, 4-22
 EW support and, 4-46, 4-64
 executing, 4-46
 ground-based assets, 4-54
 hostile collection and, 4-67
 intelligence support to, E-4
 intelligence, surveillance, and reconnaissance and, 4-45
 jamming control authority and, 5-12
 Marine tactical electronic warfare squadron capabilities, E-18

Index-2　　　　　　　　　　FM 3-36　　　　　　　　　　25 February 2009

Index

Entries are by paragraph number unless specified otherwise.

electronic attack *(continued)*
   offensive, examples of, 1-12
   Prophet and, E-7
   radio battalion (Marine Corps) and, E-10
   reporting of, 3-14
   targeting and, 4-43
   targeting working group and, 4-44
   in a time-constrained environment, in a, 4-33
electronic attack data message, D-2–D-4
electronic attack request format, D-5–D-6
electronic attack set, AN/ULQ-19(V)2, E-12
electronic deception, electronic attack and, 1-12
electronic intelligence, E-22
   defined, 1-34
electronic masking, defined, 1-39
electronic order of battle, E-18, E-19, E-22, F-6
   course of action approval and, 4-29
   EW contributions to determining enemy, 4-39
   collection plan and, 3-13
   multinational operations, for, 6-23
electronic probing, defined, 1-31
electronic protection, 1-14–1-17
   activities, 1-36–1-44
   defined, 1-14
   directed energy and, A-7
   electromagnetic spectrum, and the, 1-47
   electronic attack, compared, 1-14
   employment considerations, 4-69–4-72
   intelligence support to, E-4
   planning, F-11
   policy, 3-14
   responsibility for, 3-11
   systems development and, 1-17
   training, E-18
electronic reconnaissance, defined, 1-33
electronic spectrum management, F-10
electronic surveillance, E-18
electronic warfare. *See* EW.
electronics security, defined, 1-35

responsibility for, 3-12
electro-optical-infrared countermeasures, defined, 1-25
elements of combat power, EW support of, 2-4
emission control, 2-14
   defined, 1-41
   guidance, responsibility for, 3-15
   protection and, 1-52
enemy capabilities, evaluating from EW perspective, 4-39
E-Space, F-7
event matrix, EW contributions to, 4-40
event template, EW contributions to, 4-40
EW, defined, 1-8
EW coordination cell (joint), 6-3–6-9
   augmentation of, 5-4
   establishing, 5-4, 6-7–6-9
   EW working group as, 3-3
EW coordination cell (multinational), 6-18
EW frequency deconfliction message, D-7
EW functional matrix, 4-25
EW mission summary, D-8
EW mutual support, 6-19
EW officer, duties of, 3-12. *See also* EW working group.
EW red team support, F-8
EW requesting/tasking message, D-9
EW reprogramming, 4-74
   defined, 1-40
   multinational operations, for, 6-24
EW risks, assessing, 4-22, 4-23, 4-30. *See also* composite risk management, risk controls.
EW running estimate, 3-13, appendix C
   course of action analysis and, 4-23
   course of action comparison and, 4-26
   execution, during, 4-78
   in a time-constrained environment, 4-33
   mission analysis and, 4-14
   preparation and, 4-76
   receipt of mission and, 4-5

EW support, 1-18–1-20
   activities, 1-32–1-35
   battle damage assessment and, 4-64
   deconfliction with collection operations, 4-73
   defined, 1-18
   directed energy and, A-7
   electromagnetic spectrum and, 1-47
   electronic attack and, 4-46, 4-64
   employment considerations for, 4-73
   intelligence and, 5-15
   intelligence support to, E-4
   jamming control authority and, 5-12
   requests, joint and multinational, 4-78
   signals intelligence, compared with, 1-20, 4-73
   targeting to, 4-64
   time-constrained environment, in a, 4-33
EW systems, airborne, 4-57–4-60
   ground-based, 4-54–4-56
   testing support for, 7-3
EW training, 7-3
EW working group, assessment and, 4-81
   coordination actions of, 5-14
   course of action analysis tasks, 4-22–4-23
   course of action approval, tasks after, 4-29
   course of action comparison tasks, 4-26
   course of action development tasks, 4-16–4-20
   deconfliction and, 4-64, 5-19
   fires cell and, 3-2–3-3
   IPB and, 4-35–4-40, figure 4-6
   mission analysis tasks, 4-7–4-9, 4-12–4-14
   planning and, 4-3
   preparation tasks of, 4-76
   staff representation in, 3-2–3-3
   synchronization and, 5-20
execution, defined, 4-77
execution tasks, EW, 4-78
exploitation, detection and, 1-48

F

fires, EW synchronization and, 5-20
   integration of EW with, 3-12

Index

Entries are by paragraph number unless specified otherwise.

fires cell, EW assessment and, 4-83
 EW working group and, 3-2–3-3
fires warfighting function, EW support of, 2-13
fratricide, EW synchronization and, 5-20
 preventing, 4-42, 4-66
frequency management,
 coordination, 6-11
 plan, 6-10
 protection and, 1-52

G–H–I

G-2 staff, EW responsibilities of, 3-13
G-3 staff, EW duties of, 3-11
G-5 staff, EW assessment and, 4-82
G-6 staff, EW duties of, 3-14
Growler, E/A-18G, E-40–E-43
guarded frequencies, 3-13
Guardrail common sensor, E-5
hardening, protection and, 1-52
high-payoff targets, 4-43, 4-44, 4-45
high-speed antiradiation missiles, 4-59, E-21, E-22, E-34, E-36, E-37, E-41. *See also* antiradiation missiles.
high-value targets, EW, 3-13
 EW contributions to, 4-39, 4-43
 identifying, 4-12, 4-22, 4-23
improved capability II, EA-6B, E-36
improved capability III, EA-6B, E-37
indications and warnings, intelligence warfighting function and, 2-12, 2-14, 2-16
information engagement working group, EW synchronization and, 5-20
information operations, Joint Spectrum Center and, 7-8
 U.S. Strategic Command and, 7-6
information operations cell (joint), 3-4, 6-2, 6-3, 6-5
 multinational operations and, 6-16
information protection, defined, 2-8

EW support to, table 2-2
EW synchronization and, 5-20
functional cells concerned with, table 2-1
staff responsibilities for, table 2-1
time-constrained environment, in a, 4-33
information requirements, determining, 4-20
 EW-related, 4-43, 4-52
 initial, 4-14
information superiority, defined, 2-6
information tasks, EW support of, 2-6–2-9, 5-17
 in a time-constrained environment, 4-33
integrating processes, EW contributions to, 4-34
intelligence, electronic, defined, 1-34
intelligence activities, coordination with, 5-15
intelligence preparation of the battlefield (IPB)
 assessment during, 4-81
 course of action analysis and, 4-21
 defined, 4-35
 EW contributions to, 3-12, 4-6, 4-8, 4-35–4-40, figure 4-6
intelligence requirements, determining, 4-20
intelligence support, coordination for support from other Services, 6-13
intelligence, surveillance, and reconnaissance (ISR)
 command and control warfare and, 4-45
 electronic attack and, 4-45
 planning for, 4-51–4-52
intelligence, surveillance, and reconnaissance synchronization, 4-50–4-52
 assessment during, 4-81
 defined, 4-50
intelligence synchronization matrix, EW tasks in, 3-11
intelligence, surveillance, and reconnaissance plan
 EW tasks for, 3-11
 course of action approval and, 4-29

intelligence systems (Army), E-4–E-7
intelligence warfighting function, EW support of, 2-12
interception, radio battalion (Marine Corps) and, E-10
intrusion, electronic, disruption and degradation and, 1-51

J

jamming, E-12, E-18, E-21, E-34, E-36, E-37, E-39, E-42, E-43
 assessment and, 4-49
 degradation and, 1-49, 1-51
 disruption and, 1-51
 effects of, 4-68
 electromagnetic, defined 1-29
 electronic attack and, 1-9, 1-10, 1-12
 EW synchronization and, 5-20
 Joint Spectrum Center and, 7-8
 support to, 4-55
jamming control authority, 3-12, 4-43, 4-78, 5-12
Joint Communications Security Monitor Activity, 7-5
Joint Electronic Warfare Center, 7-7, F-8
joint force air component command, EW coordination with, 5-5
joint force EW organization, 6-2–6-14
joint frequency management office, 6-10–6-11
Joint Improvised Explosive Device Defeat Organization (JIEDDO), F-9
Joint Information Operations Warfare Command, 7-6–7-7
joint intelligence center, 6-12–6-13
joint mission planning system, E-22
joint operations, EW coordination for, 3-4
joint restricted frequency list, D-12–D-13
 jamming control authority and, 5-12
 Joint Spectrum Center and, 7-8
Joint Spectrum Center, 7-8, F-10–F-12
joint spectrum interference resolution, D-11

Index-4 FM 3-36 25 February 2009

Index

Entries are by paragraph number unless specified otherwise.

joint tactical air strike request, D-10
joint targeting coordination board, 6-14
Joint Warfare Analysis Center, 7-9

K–L–M

Knowledge and Information Fusion Exchange (KnIFE), F-13
lasers, directed energy and, A-7
leadership (element of combat power), EW support of, 2-5
lethal effects, decisionmaking example, 1-54
Marine Corps Information Technology and Network Operations Center, 7-10
Marine radio battalion. *See* radio battalion.
Marine tactical electronic warfare squadron, E-17–E-24
measures of effectiveness, developing EW, 4-82
military deception, EW synchronization and, 5-20
military decisionmaking process (MDMP), assessment during, 4-81
mission analysis, EW actions during, 4-6–4-14
mission rehearsal exercise, 4-76
mission variables, 1-2
mobile electronic warfare support system, E-13, E-14–E-16
modified combined obstacle overlay, EW contributions to, 4-38
movement and maneuver warfighting function, EW support of, 2-11
multinational operations, 3-4, 6-15–6-24
munitions effects assessment, electronic attack and, 4-48
named areas of interest, EW contributions to, 4-40

N

National Ground Intelligence Center, F-4–F-5
national intelligence, coordination for, 6-13
National Security Agency, 7-11–7-12

electronic protection and, 4-70
NATO emitter database, 6-19
network operations cell, EW synchronization and, 5-20
network operations officer, EW duties of, 3-14
nonlethal effects, decisionmaking example, 1-54

O–P

obscuration, directed energy and, A-6
offensive electronic attack, 4-61
operational environment, defined, 1-1
evaluating from EW prospective, 4-37–4-38
orders, Army, EW input for, B-1, figure B-1
joint, EW input for, B-2–B-3
orders production, EW contributions to, 4-30–4-31
particle-beam weapons, A-7
planning, assessment during, 4-81
considerations for EW, 4-1
plans, Army, EW input for, B-1, figure B-1
joint, EW input for, B-2–B-3
precipitation static, defined, A-1
preparation, 4-75–4-76
assessment during, 4-81
defined, 4-75
priority intelligence requirements, EW contributions to, 4-22, 4-39
Prophet, E-7
protection, 1-52
protection warfighting function, EW support of, 2-16
Prowler. *See* EA-6B Prowler.

Q–R

Q-19(V)2 electronic attack set, E-12
Q-36 mobile electronic warfare support system, E-13
Q-36A mobile electronic warfare support system, E-14
Q-90 tactical electronic reconnaissance processing and evaluation system (TERPES), E-19, E-23–E-24
Q-99 tactical jamming system, E-21, E-34

Q-113 radio countermeasures set, E-34, E-39
radio battalion (RADBN), E-9–E-16
radio direction finding, radio battalion and, E-10
radio frequency countermeasures, defined, 1-26
radio reconnaissance teams (Marine Corps), E-10
radio-frequency weapons, A-7
RC-135V/W Rivet Joint, E-29–E-31
receipt of mission, EW actions on, 4-4–4-5
reconnaissance, electronic, 1-33
red team, electronic attack planning and, 4-67
support, F-8
rehearsals, EW support, of, 4-76
reprogramming, 5-16
Joint Staff oversight of, 7-7
restricted frequency list, 3-13, 3-15
risk controls, 4-30. *See also* composite risk management, EW risks.
Rivet Joint, RC-135V/W, E-29–E-31
rules of engagement, EW, 3-12
running estimate. *See* EW running estimate.

S

S-2 staff, EW duties of, 3-13
S-3 staff, EW duties of, 3-11
S-5 staff, EW assessment and, 4-82
S-6 staff, EW duties of, 3-14
sequels, 4-22, 4-23, 4-75
signal operating instructions, 3-15
signals intelligence, E-4, E-13, E-14, E-16, F-7
aerial common sensor and, E-6
assessing electronic attack and, 4-49
EW support, compared with, 1-20, 4-73
EW-related information requirements and, 4-52
foreign, 7-11
Guardrail common sensor, E-5
multinational operations, for, 6-22

25 February 2009   FM 3-36   Index-5

Index

Entries are by paragraph number unless specified otherwise.

signals intelligence *(continued)*
  preparation and, 4-76
  Prophet and, E-7
  radio battalion (Marine Corps) and, E-10
  support requests and, 5-8
  support to battle damage assessment, 4-64
  support to targeting, 4-43, 4-45, 4-64
situation template, EW contributions to, 4-40
situational awareness, electromagnetic spectrum, of, 2-16
special compartmented information facility (SCIF), joint EW coordination cell and, 6-8
special intelligence communication, radio battalion (Marine Corps) and, E-10
special technical capabilities, determining requirements for, 4-18
special technical operations, joint EW coordination cell and, 6-8, 6-9
spectrum management, 5-9–5-11, F-10–F-12
  electronic protection and, 1-16
spectrum management plan, electronic attack and, 4-46
spectrum manager, duties of, 3-15
spectrum supremacy, 7-8
support requests, 5-6, 5-7–5-8
suppression of enemy air defenses, 1-12, 2-11, 2-16, 4-59, E-34, E-42
  execution, during, 4-78
  jamming and, 4-68
sustainment warfighting function, EW support of, 2-14
synchronization matrix, EW contributions to, 4-22, 4-23, table 4-1

T

tactical electronic reconnaissance processing and evaluation system (TERPES), E-19, E-23–E-24
tactical jamming system, AN/ALQ-99, E-21, E-34
target areas of interest, EW contributions to, 4-40

targeting (process), 2-13
  assessment during, 4-81
  defined, 4-41
  detection and, 1-48
  EW integration into, 3-3
  EW support of, 2-12, 2-13, 3-7, 4-41–4-49, 4-64, 4-73
  joint, 6-14
  preparation and, 4-76
  signals intelligence support of, 4-64
targeting information, Guardrail common sensor and, E-5
targeting working group, electronic attack and, 4-44
time-constrained environment, EW working group decisionmaking tasks, 4-32–4-33

U–V

U.S. Strategic Command, 7-6
USQ-113 radio countermeasures set, E-34, E-39
visualization, EW employment, of, 2-3
VMAQ. *See* Marine tactical EW squadron.
vulnerabilities, EW, 3-13
vulnerability analysis and assessment, 4-70, 7-4

W–X–Y–Z

war-gaming. *See* course of action analysis.
wartime reserve modes, defined, 1-43
working group, defined, 3-2. *See also* command and control working group, EW working group, information engagement working group, targeting working group.

Index-6 FM 3-36 25 February 2009

Printed in Great Britain
by Amazon